Ninus J. Piter

"Quantum Physics" Wonderful Story

From Planck to Aspect
Physicists Are Hearts and Brains

Translation by Ninus J. Piter

themindedge@gmail.com - https://themindedge.blogspot.com/

© 2021 Ninus J. Piter

Original Title

Einstein, Mia Nonna e La Fisica Quantistica

https://www.amazon.it/dp/B098WDGQ1N

© 2021 Printed by Amazon Fullfillment

ISBN 9798457890886

Cover Illustration: © Ninus J. Piter

First edition 7 July 2021

First digital edition 7 July 2021

This work is protected by the copyright law.

Any unauthorized duplication, even partial, is prohibited.

Reserved literary property.

Presentation

Each reader is invited to read this book in a different way, according to the path he will want for an independently trace: following the chronological thread of the adventure from Planck to the recent confirmation of the Odderon and the discovery of the Muon G2; or by choosing the chapter casually to read and building the plot himself; or by investigating the scientists who built Quantum Physics and closely following the men to whom this book is dedicated.

The experiments, the obstacles, the intuitions will be investigated and assimilated with different interpretation compared to what has been done. Until discovering a new way of understanding this branch of science so as to transform himself into the protagonist of a story yet to be written, if only he will have the desire.

The author's hope is that readers, after reading up to the last line, will see the world with new eyes and decide without any conditioning what is the reality in which he lives.

Dear reader,

before reading the adventure that this book offers you, you need to know a few things: I translated this completely by myself. So I apologize in advance for any translation error I may have made. I invite you to report them to me but, above all, I invite you to continue reading because what I have to tell and demonstrate is much more important than the small mistake you may have come across.

If there were parts that you think are too technical, I invite you to skip them: the story I hope to tell you will not be affected. Eventually you can read that parts later.

My email address is:

themindedge@gmail.com

Please feel free to contact me and report any errors: as far as possible I will try to answer. Enjoy the reading.

Thanks

Ninus J. Piter

In what I am about to write, I judge myself, that perhaps will say nonsense, but also judge the pig headed opponent who always denies blindly:

"Here, know-all, what do you speak about?"

Knowledge is not only in the head of that Old Minded Teachers aligned along the street of intransigence.

As "Quantum Physics" teaches, life is here and elsewhere at the same time, the cat is alive and dead at the same time.

So in the time when I say that someone has wrong, I am also the other, which suspects to have said a truth.

Anonymous Quote

If you're not failing every now and again, it's a sign you're not doing anything very innovative.

Woody Allen

My Grandma

I like to remember of my ancestors and I have so many pleasant memories that, although hidden somewhere in my mind, just investigated they jump out. These memories have put roots because there was always with me a great storyteller, my grandmother, who told me stories whose performers were often the result of her imagination or old members of my family.

I will never know which of these stories were true. Surely I know that in this way, today, I know a lot of my ancestors.

My grandmother's name was Constantina and was born on October 1, 1884. In those same years Joseph John Thomson proposed the first atomic model called *Plum Pudding Model*.

I always think that, while my grandmother challenged life, "Quantum Physics" grew and developed hand in hand, as if it had the same rhythms and the same difficulties of a child, then a young boy, who became a teenager ready to fight against the world and, at the end, to become an adult with its maturity and awareness.

Finally, in a natural development, the time comes to give way to a future heir who will recognize merits and expense life, because the descendants can progress the path started, towards new goals.

My grandmother had never heard of Albert Einstein's *Theory of Special Relativity* and, at the time of the first publication of the article, she already was twenty-one years old.

She was the firstborn of a humble family and the most important task she had was the duty to pay attention to the other younger brothers and sisters and was a great thing that, completely self-taught, she learned to read, write and take into simple mathematics. Southern Italy of that period was so concreteness in the daily life to grant little room to the news of theoretical scientific discoveries.

When in 1927 he got married, unknown men to her decided the atomic future of humanity, meeting in the conference of Como and in the Congress of Solvay.

My grandmother, on the other hand, the next day her marriage returned to her duties of woman and wife to contribute in her own way to the growth of a humanity that fought to make progress along the way of daily social and scientific progress.

In 1945, when the two atomic bombs exploded, as the concrete application of the results of young "Quantum Physics", my grandmother understood only that so many people had died in a single bombardment and thought of all those little lifeless as she cared for her boys almost in age for marriages.

She had found love without knowing the Dirac's equation and the atoms continued their probable existence in bread that she weekly kneaded.

When my grandmother died, in 1979, "Quantum Physics" was now, itself adult too.

And it was probably already dead or dying.

I have always wondered what modifications have the atoms of the body of a person who dies, if they are in a self awareness, or they are just part of a great tourbillon, where no one has the real consciousness of what we are or that the true purpose is nothing else that the perpetuation of a memory.

It seems to me that knowledge and theories have the same destiny to dissolve as shadows in the light of new discoveries.

It's my opinion that both, my grandmother and all those who collaborated in the growth of "Quantum Physics", have contributed in producing results that will be perpetuated for generations to come, however and forever, no longer as manual work or through daily experiments, but only as intangible memories that will evolve into scientific stories or theories.

In the years far to come, the theories and the stories of each of the protagonists, past or present, will become *Memes* and, for our future descendants, it will be increasingly difficult to distinguish whether there really existed a reality that we believe to be models, or if we have fulfilled the sole, unconscious task of being a medium, from generation to generation, for the oldest *Memes*, perhaps enriching the multitude even with experiences or facts acquired in the course of our life.

So I hope to be able to convey both, the memory, the Meme, of my grandmother and the memory of "my" "Quantum Physics" and how much I was passionate about it.

We Were Only Men ...

Einstein was sure that

You do not really understand something unless you can explain it to your grandmother.

... and so I tried to explain "Quantum Physics" to my grandmother while she was cooking for me.

A branch of physical sciences that fascinates and awakens a sense of wonder, curiosity and astonishment at the mere pronouncement of its name, is "Quantum Theory" with its mathematical representation that describes its behaviours, called "Quantum Mechanics". Neophytes and all those who approach them unencumbered because of the lack of knowledge they have, but recalls great expectations, enthusiasm, that as you begin to understand they turn into perplexity, uncertainty, scepticism until you create doubts that this theory may even be false.

All of this is the same approach as its creators, who could not fully understand it. And, in my opinion many scientists, who are still working on it today, do not fully understand what it conveys to us, despite the more than a hundred years spent checking, interpreting, predicting or trying to visualize what has evolved since its birth.

The result of so many fundamental contributions has produced such a variety of interpretations that, instead of making it simpler, it increasingly complicates the understanding of the composition and structure of matter at the atomic and subatomic levels, always reserving new surprises.

Let's take a walk together on the sand of a beautiful exotic beach listening to the rising waves and brush the sand with our hand and look at a tiny grain of sand that has remained attached: that grain of sand is composed of a number of atoms similar to the total number of sand grains of the beaches of the whole world!

The sound of the undertow spreads in the air and, through sound waves, reaches our ears, traveling exactly like the waves associated with elementary particles.

Down there, in the most intimate part of matter, we discovered that there was so much space and so many components that we could not even imagine with the wildest imagination. Not only that, "Quantum Mechanics" has pointed out that there are laws that subatomic elements have to undergo, completely different from those we're used to. Read that, inevitably, while reflecting on the world we live in every day, we would never have been able to identify without this branch of physics. Readings that are very different, more complex, often hidden from macroscopic events and, so complicated and enigmatic are still largely impenetrable to the human intellect.

This privilege has been granted to us only in recent times. Now we live with a universe that was unknown to us and has very few to share with the classical physics that explained the world around us.

"Quantum Physics" appears as a theory that is not very intuitive, illogical and, at first glance, so senseless that it astounds its own multiple fathers.

In the 1965's Richard Feynman, one of the best minds in physics in the second half of the 20th century, with his irony seemed to comment:

> *I think I can safely say that no one in the world really understands "Quantum Mechanics".*

Let us eliminate all doubts, contradictions and remorse. Let us not be fooled by *New Age or Exotic Theories*: this beautiful and very complicated physical theory works perfectly, but only thanks to a very sophisticated mathematics. The physical-mathematical combination is indivisible and there are no other possible interpretations: this is only science. The findings have been based on technological innovation, which has produced applications and tools used on a daily basis, including all laser and nanotechnology.

Not only that, biology and medicine are progressing more than ever thanks to the great results achieved.

The comparison, now becoming daily and fascinating, involves not only scientists of different opinions, but also philosophers, sociologist's writers, not to mention many movies interpretations.

My desire is to make our grandmother fall in love and understand the secrets of "Quantum Physics", too, with a chronological narrative that avoids the use of sterile mathematical formalism and, that is reconstructed on the story of men and events that generated it.

An evolution often caused by random factors, sometimes stubbornly sought through the examination of experimental results, by lonely men or associated in research groups.

Everyone has contributed to this by sharing the results.

Our senses, limited from human nature, are not suitable to have full awareness of the environment around us: our eyes do not see the colours of ultraviolet or infrared, our hearing allows us to hear, in the best case, only the frequencies between 20-20000 Hz. However Einstein argued that we have a gift that is a higher quality than intelligence itself: the ability to imagine.

Our remarkable intuition has always allowed us to deduce complex theories by linking individual elements seemingly insignificant and detached from the context. Thus we have compensated these innate deficiencies with the acuity of the ingenuity and with the ability to visualize, through the formulation of a theoretical mathematics, models that have allowed us to "evolve" our understanding capacities.

The evidence of classical physics is thus replaced by the "strange" conclusions of "Quantum Physics", as Feynman has pointed out on several occasions. The same scientist often invited his students not to approach the studies with a very narrow logic but to let them be convinced by intuitions rather than by traditional models.

For him it was not necessary to ask *how it could be so?*, otherwise we would have entered *a blind alley from which no one has yet come out*.

These scientists were and still are men first and foremost children of their own time and of their own societies. They have the urges and weaknesses common to all of us, the environments in which they express themselves and which have made them progress. They are the same ones in which we live and their teachers, their games and all the coincidences of their lives are the same we know.

It is as if everything had worked to achieve a common goal for all that is to advance them as explorers of all humanity in the new fields of knowledge.

Behind every scientist are all of us and every achievement of the individual is in common with everyone. We are a single interconnected organism.

And so it is for the whole universe, as we can see by continuing reading.

Forever and ever, at any time and anywhere.

The Universe

To understand physics and all the events that affect it, one must have a very clear idea of what the Universe is.

The term originates from the Latin word *Universum*.

The meaning of the term, contrary to the widespread use, is complex and determined by the ways of thinking of men, adapted to the physical knowledge and philosophical interpretations of each period in which it is used.

By trying to explain it exhaustively, we could say that the term indicates what man wants to "see", leaving a very subjective and adaptable meaning to the knowledge of each and every time, past, present and future.

Chronologically, from the origins of human thought, we first must include *everything*, i.e. both the content and the container, of all things that exist, everywhere, from micro to macrocosm, including all phenomena that occur in this all one.

Then, looking up at the cosmos, we identified it as the globality of the whole we could see in space, including our Earth.

It was implicitly assigned the perfection of the *Initial Will* for which everything exists and which corresponded to the natural order of everything understandable, that is the natural order of all things that we see, understand or that remains hidden because we do not understand it.

In the Middle Ages the universe expanded to understand both creation and non-creation, thus becoming an expression of the will of a creation.

For us, men of the 20th century, it has become space-time, the movement of bodies, matter and physical laws and all models that can explain natural phenomena.

Following the formulation of the *Theory of General Relativity*, new boundaries have been imposed that we will never know, because it assumes that, because of the displacement of certain regions of the cosmos at a faster rate than that of light, they will never again be able to interact with our physical experience, but, especially we will never know: we could also say that there is nothing else even though we know that there is something else.

That is why today, because of this way of thinking it, we no longer speak of a general universe but give it a physical limitation by defining it as the *Observable Universe*.

In the Universe there are billions of galaxies and thousands of billions of stars around which they rotate an unimaginable number of planets, in addition to moons, asteroids, comets and clouds of dust and gas: all bodies that move in the Universe obeying laws established immediately after the initial moment.

All these bodies are made up of atoms, especially hydrogen, which is the most common element in the Universe, and helium. And that's only about five per cent, because everything else we can't determine yet.

Once, a person explained to me the Universe as the Latin meaning *Uni-Versum*, something that goes all together, with a unique principle, in the same direction.

Something that always strive towards one goal.

For all those who are believers in a Creator being, therefore, the Universe could be understood as the whole, visible or invisible, known or unknown, understandable or yet to be discovered, which strive towards one only goal: its Creator.

Theory Satisfaction

In according to scientists, the concept of "theory" is different from how most people understand it. In common language it indicates an abstract idea, in scientific language it indicates how we interpret the facts by giving them a logical-mathematical construct.

The original etymology of this word take origin from the Greek θεωρέω (theoréo) that means "*I watch, I observe*", and it is the combination of the two words, θέα (thèa), "*show*" and ὁράω (horào), "*watch*".

Thus a "theory satisfaction" is the capability to "watch and to observe a show".

When a theory is confirmed, it is like to watch a complete show in its totality of plot, script and dialogue.

For a scientist, or a team of them, to "watch" proven hypotheses formulated in a working theory is not only a success but a real stepping stone to overcome the next obstacle and propose new goals. After fulfilling a theory, you can only go on; the present is already past and has begun, without a solution of continuity, a new path to satisfy another theory.

It is very important that we do not forget this way of thinking otherwise we would be stuck trying to embellish what we have already done, fiddling without innovation. A bit like in the Baroque: we worked only on the past making it more and more beautiful, but preferring it to the charm of the suspense of the future.

Innovators cannot be afraid of anything the future holds for them.

This is a real danger for "Quantum Physics": we'll find out where it came from and if it found the new trampoline to have new incentives for the next jump.

All scientific theories are based on hypotheses: solutions are suggested to explain that new phenomenon not yet understood and not explained from any available scientific theory. The hypothesis is an idea not yet proven; the theory is the transformation into a scientific model of the enunciation of hypotheses.

Let us take an example: a theory is like a bottle containing some liquid. The liquid is the totality of all the observations and ideas that scientists have about an experiment. This liquid follows the shape of the bottle and, therefore, even with the same content, it is the bottle that determines its perception. The bottles thus represent theories that can offer a different visibility even if containing the same elements.

It is also true that two identical bottles may contain different liquids; so the same theory could explain completely different data from the same experiment.

Theories must take into account all assumptions made and must be concise in the exposure, consistent with observations, repeatable and predictive. From a scientific theory that satisfies a large number of hypotheses it is possible to extrapolate a model or a set of models that define a general show, usable as trace and prediction, for any other data or detection carried out respecting the theory.

By convention, a theory is never fully proven, because we cannot know everything about the phenomenon to which it is applied. In the scientific method theories and laws are two distinct things: the law describes a phenomenon of nature stating that it is true without clarifying why, while the theory explains the observations of that phenomenon coordinating them all together.

Satisfying a theory is the ability of the observations of an experiment or event to adapt to that theory: the results confirm the hypotheses and since then all phenomena of the same type, already observed or which will be in the future, satisfy the theory generated by the hypotheses. A satisfied theory is also predictive because we know even before we make a measurement that, within the same phenomenology, the experimental results we will obtain will be explained by the theory. For example, the motion of every object that falls is explained by Newton's Theory, every living being responds to Darwin's theory of evolution.

The theories are valid until proven otherwise: in fact, once formulated there may be a need to adapt and modify it according to the updated observations, but it must remain valid until other observations of the same phenomenon find a better explanation.

The guarantee of its accuracy also serves the opposite, as valid instruments and experiments made in other places or times.

This instrumental and intellectual efficiency grants the scientificity and the ability to objectively judge one's own knowledge, without the slightest doubt of the authoritativeness of the basic theory.

It is also necessary to consider those experiments which produce results or observations without these being hypothesized or researched. It is the *serendipity*[1] of science: theories that seem to

---

[1] *serendipity*, faculty for making discoveries by accident, not necessarily what you was looking for.

have been given to us by a case at that time benevolent to us and that work perfectly. In any case, the researcher's mind must always be ready to detect and interpret the phenomenon correctly and then propose the appropriate theory.

Atoms

> *It would be a poor thing to be an atom in a universe without physicists, and physicists are made of atoms. A physicist is an atom's way of knowing about atoms.*
>
> *George Wald*

First Atom Definition

> *Perhaps twenty-five centuries ago, on the shores of the divine sea, where the song of the aedi had just died out, some philosopher was already teaching that changing matter is made of indestructible grains in continuous movement, atoms that chance and fate would have gathered in the course of centuries according to the shapes and bodies that are familiar to us.*
>
> *Jean Perrin (1870-1942)*

Jean Perrin (1870-1942), French chemist and physicist told, with due caution because to the few sources available, that the first in the history of humanity to conceive the idea of atom were Leucippus of Miletus, belonging to the Eleatic school and probable disciple of Parmenides and Zeno and Democritus of Abdera (5th century BC), his disciple and friend, who will give rise to the philosophical school of Abdera. Not much is known about the history of their lives or about their works because they were lost or destroyed, but contemporary sources agree that this intuition is to be attributed to them.

In the *Makros Diàkosmos, Great Cosmology* of Leucippus and in the *Mikròs Diàkosmos, Little Cosmology* of Democritus, their atomistic ideas about nature and the cosmos were exposed.

Leucippus and Democritus, followers of Eleatic philosophy and, indirectly, of Pythagoreans, developed the idea dear to the Pythagoreans, according to which the phenomenology of the physical world can be revealed with mathematical explanations. Eleatic philosophy formed them at the Parmenidean principle of being and non-being: innate, indestructible, coherent in its parts and unalterable, without an end to aim for, no beginning or future, since it is a homogeneous, unique and continuous whole.

> *Being is and cannot not be, not being is not and cannot be.*
>
> *Parmenides, 5th century B.C.*

The atomists modified the principle by adapting it to their thought: the *being* becomes multiple, so there are more individual entities with shape and size and in continuous motion. This multiplicity and their movement force us to reconsider the *not-being* that becomes appearance and opposes truth with the concept of *emptiness*. Institutions become an infinite number, imperfect and indefinite.

These entities, so called atoms, in their multiplicity are invisible and indivisible, because if they were visible they would also be divisible. Their dimensions, shape, arrangement, direction and position are peculiar characteristics of each and each characteristic has qualities, so they can be smooth, spherical, sharp, rough, curved or crocheted. In their eternal disorderly motion, they collide, divide, compact forming bodies.

Anassagora, contemporary philosopher of Democrito, assumed that there were even smaller parts to form the atoms, such as clay that when compressed allows to realize the bricks, which in turn become elements of a wall.

For the total causality of the atomic movement, Dante assigned Democritus to Hell, because he believed the world was derived from the case without the perfection of a divine Creature.

Our present society is essentially based on the influences that derive from Greek culture, whose way of understanding and combining mathematics, logic, philosophy has inevitably and unconsciously contributed to influencing the knowledge and the method we use to understand the experience of the surrounding world, combining them in a constructive synergy with our intelligence. That is why we cannot define this process "rational" or "not rational", because there is no objective observer to judge a truth.

For this reason Epicurus, who was indirectly influenced by Abdera's school because his teacher Nausifane belonged to it, accepted Democritus's idea that atoms were always moving in space, but added that, because of their weight, they fell to the ground because heavier than the vacuum surrounding them involving the case, which, by deviating the trajectory makes them collide allowing the aggregations that form the matter.

Epicurus amended Democritus's idea of atoms with infinite forms to a numerable number, but so great that it could not be imagined.

Many of Epicurus's writings have also been lost, but we can find many of his thoughts reported in the *De rerum natura* of *Titus Lucretius Caro (98-55 a.C.)*, a masterpiece of Latin literature. Six books in verses designed to convince Roman politician Gaius Memmius and dedicated to him, in which Titus Lucretius Caro described the nature of the world and of men and the materialist conception of atom was the pivotal concept of all the work. The arguments he used were the consequence of a subtle mind, mostly if we refer to the period in which he was drafted. Lucretius supported the idea that atoms were the *basic bodies* to be subjected to wind modelling that used *invisible bodies* in his blowing. The *flavour*s were invisible particles reaching our nostrils and, finally, the definitive proof of the existence of atoms was the effect that in proximity to the sea first moistens and then dries the tissues if exposed to the rays of the sun. Lucretius also justified micro-events, such as the growth and wear of things, with the atomic hypothesis.

The work of Lucretius has an original poetic beauty and after twenty centuries encloses an astonishing correctness. The nature observed rationally produces almost a calming effect on the fears and anguish of man caused by superstition.

It may seem strange, but since then there was a scientific globalisation because, even in the rest of the known world, other cultures have asked themselves the same questions about the composition of bodies and matter.

Indian doctrine of *Jaina* (the *Victorious* appellation given to asceticism that founded it in the 5th century BC) tried to explain the concept of atom and in the 4th century BC, the members of the current *Hinayana*, one of the two fundamental doctrines of *Buddhism,* still existing today, expressed similar atomic ideas.

Due, however, to the "holistic"[2] vision of reference, for which reality is a whole and intimately connected, Oriental people ideas never developed comprehensively with atomism.

The idea of Democritus and Leucippus of *Atom*[3], perfected by Epicurus and Lucretius, was to

> *Simple particles (invisible, unalterable and indestructible) of which every substance is composed, which, moving in the void, colliding with each other and composing in various ways (in fact, they were imagined with hooks and protuberances), give way to things as they appear.*

These eternal particles were the bricks, the elements, which by combining randomly constituted all the matter known.

The atomic concept, although questioned on more than one occasion, was never set aside because it satisfied the explanations of the elementary characteristics of matter in its three states, solid, liquid and airborne.

It was the chemists at the beginning of the 19th century who began the work of changing atomic ideas with their experiments.

---

[2] *Holism* sees the whole as the sum of its parts. The sum of the parts is always greater than one.
[3] *Atom* comes from the Greek and its meaning is *indivisible*, because it could not be divided.

Infinite Idea's, Origin and End of All

In the last pages we introduced a new concept, the *Infinite* and, we accepted it calmly assuming that we knew what it was.

However, would we really be able to explain what it is, or do we flatter ourselves on convictions deriving from its frequent use without understanding what we are talking about?

Yes, this is really important because we will find the concept expressed many times.

Discussion about *Infinite* is a vain feat: we will never be able to agree on either the definition or the concept it expresses. There is the *infinitely* small and the *infinitely* large. And at the end, they might even coincide.

One thing we can be sure of: the *Infinite*, from the last years of the 19th century, has assumed a primary role in scientific consideration.

In ancient times no one has ever had problems with the *Infinite*, neither the Egyptians and Babylonians, nor the Mayans or other Eastern cultures: it was simply a problem that did not arise. Everyone, in their pragmatism, saw a world limited to their real perceptions.

The first to conceptualize the infinite mathematical and philosophical were the Greeks, but they considered it in a very different way from us.

Euclidean geometry is much more valid than we have commonly been led to believe, and the concepts it expresses are still valid even when it comes to "Quantum Physics".

Some definitions have to be traced back to the original explanations so that they can be understood in the light of the progress made: first of all, we have to remember that the point is dimensionless, and that is exactly how it is still valid today.

A point has no parts, no length and no width.

Even the postulate on the geometric line is not as we pronounce it. The Euclidean version says that *every line can be indefinitely extended*. The concept of infinite is not mentioned in any way.

And even the famous *parallel line* does not meet at one point: because of the indefiniteness and the difficult understanding of the concept of infinity, in the historical period in which the original formulation was pronounced spoke of a lack of points in common. *Two parallel lines simply have no points in common.* Euclid preferred not to quote the Infinite in order not to find himself in conceptual difficulties and was very careful not to speak of beginning or end.

In summary, the early Greek mathematicians did not mention explicitly the *Infinite*, the *Origin*, or the *End* for any basic geometric entities, but let their potential be intuited by omitting any verbal description. In fact, they were even more modern: "they were able to increase quantities with finite values indefinitely".

It is a much more modern concept than what we have been used to saying in our school lessons that raises many doubts about how this knowledge is conveyed and, over time, used by our students.

Anaximander that was a contemporary politician and mathematician with Thales, in the 6th century. B.C. speaks about the *archè*, the *Origin of all things*. And the principle of all things, for the first time, is not identified in the four natural elements, *air, water earth and fire*, but in the *ápeiron*, a kind of infinite/indefinite principle from which everything derives and to which everything will move. According to Anaximander, the concept derives from an initial *injustice* that has led to the separation of the perfection of all in its opposite "hot/cold", "wet/dry", etc. The *ápeiron* has divine connotations because it is immortal, indestructible and permeates and regulates the universe. How many similarities with what will become the research of the century: the space-time opposition, Matter-Antimatter and theories on the Aether, better known as quintessence or fifth element.

The *ápeiron* (from the Greek ἄπειρος or ἀπείρων, is composed of ἀ-, "not", e πεῖραρ, *peirar*, "limit" o "end") is everything limitless, which has neither a small nor large space limit.

John Wallis, an English mathematician in the 1655, uses for the first time the symbol of infinity as we know it today: ∞.

In mathematics and physics it refers to a quantity without limit or end, greater than any perspicuous number or size.

Atomic Theory Is a Good Idea

During Middle Ages and Renaissance, the only possible scientists were alchemists and science was seen as mysticism and mystery, except in a few cases. But these were the first steps for important future developments: alchemy will be the origin of the chemistry that will study atoms.

However until the twelfth century, rarely somebody talked of atoms, but rather of corpuscles of matter.

Robert Boyle (Lismore, 25th January 1627-London, 30-December 1691), an Irish chemist, physicist and philosopher, in 1662 proposed the hypothesis that gases are composed of so many moving microscopic particles that everyone collide with each other when pressure, volume and temperature change.

These particles cannot yet be called atoms because, from 1623, any reference to atomic theory was banned in all Jesuit schools and considered almost heresy by the Catholic Church.

Antoine Laurent Lavoisier in 1775 and Joseph Louis Proust in 1798 continued the path of science and formulated the two laws that will become fundamental to John Dalton's atomic theory. Lavoisier, for the first time, spoke of the law of mass conservation:

> *Nothing is created and nothing is destroyed, everything is transformed*

and Joseph Proust affirmed the law of definite proportions:

Two or more elements that react to form a given compound always combine according to defined and constant proportions in mass.

Thus the two scientists became the fathers of modern chemistry by explaining the distinction between elements and compounds and the rules that still establish the exact weight ratio between the quantities of the elements in order to take place the chemical reaction.

1800 First Atomic Theory of Matter

John Dalton (Eaglesfield, September 6, 1766 – Manchester, July 27, 1844), an English chemist and physicist, in the early years of the nineteenth century, resumed the philosophical ideas of Democritus giving them a scientific background, from the two fundamental laws previously proposed by Lavoisier and Proust. And so he elaborated the first atomic theory of matter:

- *atoms are very small particles that constitute all the elements;*

- *all the atoms of the same element are equal to each other and they have the same mass; atoms of different elements have different mass;*

- *the atoms of any chemical reaction retain their characteristics; atoms join together or divide, and neither is created or destroyed, and are not converted into different atom. This confirmed the law of conservation of mass;*

- *atoms of different elements combine to form compounds. In a compound the atoms join in constant ratios, confirming the law of definite proportions: if the ratio between the atoms of a compound is fixed, the ratio of their respective masses will also be fixed...*

The existence of atoms explained all the experimental laws until then known, derived by the observations of natural phenomena. So the chemists, who were the first to adopt Dalton's theories, moved on to discuss the differences between atoms and molecules and how they should be named.

Physicists, on the other hand, did not favourably accept these new ideas on atomic theory: it would have been time and experimental tests to break down the resistances and induce them to retrace their steps, convincing them of the existence of atoms and of the need to deepen this new concept.

1880 Atomic Structure Time

In the second half of the 19th century the whole world was in ferment. Physics and chemistry had begun to run, searching more and more for answers to questions that were previously considered irreverent and insoluble.

Joseph John Thomson (Manchester, 18December 1856- Cambridge, 30-August 1940), a British physicist, in the 1880's was just twenty-four years old when passed brilliantly the very tough exams to get the bachelor's degree in mathematics. He became a Fellow at Trinity College and received a scholarship to continue his research on mathematical models to be applied to electromagnetic forces.

In 1884 he was appointed Professor of Experimental Physics and Director of Cavendish Laboratory to replace Lord Rayleigh and founder J. C Maxwell!

Since he was not initially an experimenter, he used his constancy and genius to gain a great deal of experience that enabled him to make *Cavendish* the European reference point in experimental physics aimed him to deep understanding the electromagnetic phenomena and the interpretation of more and more experimental data produced by experiments related to atomic physics. His colleagues and students achieved seven Nobel Prize under his leadership.

Thomson was directly interested in coordinating and checking the daily activities of his students and researchers. Under his leadership, the young scholar Ernest Rutherford began his studies to become his heir, but above all to become a future pillar of nuclear physics.

Jean Baptiste Perrin, in 1895, was an assistant at the Normal Superieure School in Paris and while studying cathode rays was able to understand that they carried negative electric charge. It was the first revelation in modern physics of the idea of electrons. His later studies undoubtedly allowed us to have evidence that atoms exist and that they are aggregated into molecules, of which he was able to determine experimentally, the dimensions.

In the 1896 Thomson, just returned from a tour of conferences and lectures held in the United States at Princeton, deepened his experiments on the negative charge found by Perrin, conceived experiments about the structure of atoms and the causes of electromagnetism. They brilliantly allowed him, working with cathode rays, to intercept and characterize for the first time the negative charge as property of a new element, a particle that will be called electron some times after.

The 30th of April 1897 he announced this discovery at the Royal Institution.

Crookes, Hertz and Lenard confirmed the new discovery with their experiments which immediately raised questions about its nature as a particle of matter or as an electromagnetic wave.

Thomson improved the experimental phase and with subsequent studies succeeded in verifying how cathode rays are made of corpuscles and that these belong to atoms.

Today talking about Thomson's experiments and conclusions makes us almost smile: we are now accustomed to the idea that there are particles called electrons, but in those years the announcement of the existence of a subatomic corpuscle must have seemed, to the scientific community, something crazy.

Thomson used a Crookes tube for his experiments: it is a durable glass tube in which there is made the vacuum and then it is filled with the rare gas that you want to study. Inside, at the top, two metal electrodes welded together are connected to a current generator. The positive pole (anode) is usually composed of an aluminium disc and the negative pole (cathode) is smaller.

When the generator is switched on, the rays starting from the cathode go towards the anode and hit the bottom of the tube generating a light zone due to the excitement of the glass atoms, called *fluorescence effect*. If an obstacle is placed within the tube between the two electrodes, usually a metal-lined cross or a mill-type wheel, the shape of the obstacle is drawn on the fluorescent area or the wheel shall start to rotate. This shows that the cathode rays, travelling in a straight line, are blocked by the metal ant and have a mass that, by hitting the small blades of the wheel, makes them move.

This experiment allowed Thomson to formulate the following hypothesis:

> *I see no other possibility except that cathode rays are charged with negative electricity carried by particles of matter. But what are these particles? Atoms, molecules or even more subtle states of matter?*

Later, with a third experiment, he applied a magnetic field around the tubes and measured the deviation curve that the previous rays underwent, and so he was able to measure the charge / mass ratio.

The ratio was more than a thousand times lower that of the hydrogen atom and allowed him to conclude:

> *We have matter in a new state...*
> *and all chemical elements are so*
> *built.*

These new ideas seemed so absurd that most of his colleagues, obviously, refused to accept them, not understanding the experiments and believing that Thomson was making fun of the whole scientific community.

But George Francis FitzGerald[4] (Dublin, August 3, 1851 - Dublin, February 2, 1901), an Irish physicist member of the Royal Society of London, confirmed Thomson's insights by interpreting particles as free electrons.

In 1891, G. J. Stoney baptized the particle found with the name *electron*: this word identified the unit of electric charge. Later, Joseph Larmor used this word in his electromagnetic theory and so it became in common use. The newly discovered electrons were not yet seen as an integral part of atoms.

> *Electron:*   *is from the Greek word ήλεκτρον (pronounced électron), amber. In the seventh century BC the Greeks had already observed that by rubbing amber or ebonite objects*

---

[4] FitzGerald is also remembered for the conjecture that bears his name, according to which all moving objects "shorten", contracting in the direction of their motion. Hendrik Antoon Lorentz, a Dutch physicist, in 1892 also came to a similar idea by applying it to his own theory of electrons. This conjecture, called the FitzGerald-Lorentz contraction in 1905 became an important part of the Theory of Special Relativity.

with a woollen cloth, these attracted small corpuscles of matter. In the sixteenth century William Gilbert, observing the same behaviour for other materials and begins to use the terms "electric force" and "electrified" for phenomena of this type. Until Stoney formalizes the name by baptizing the particle discovered by Thomson:

"... An estimate was made of the actual amount of this very remarkable fundamental unit of electricity, for which since then I dare to suggest the name of electron"

George Stoney

Further experiments with cathode rays confirmed that the observed particle had a charge and it was negative. This automatically assigned the atom a neutral value of electric charge since nothing is detectable at rest without any stress. In the 1897's these ideas convinced Thomson that, in order to maintain its neutral charge, in addition to the negative electron, the atom must have other parts that cancel it electrically. And so he generalized the concept that all atoms, in a normally neutral nature, had components with negative charge and other particles, still unknown, with positive charge.

Henri Becquerel and his wife Curie, in the 1896, had luckily discovered the radioactivity of some elements originally present in nature that had themselves corpuscular or electromagnetic or both peculiarities at the same time.

Usually all the great scientific discoveries are the result of many small steps and of the contribution of talented scientists who work in the background without receiving the recognition they deserve.

This was not the case of Antoine Henri Becquerel (Paris, 15th December 1852-Le Croisic, 25th August 1908), a French physicist who, thanks to *luck* and *serendipity* (this term is used to indicate an event entirely due to the case) made the initial discovery of radioactivity for which he received the Nobel.

Wilhelm Conrad Röntgen, (Lennep, 27th of March 1845-Munich, 10th of February 1923), was a German physicist that discovered in 1895 the X-rays for which received the first Nobel Prize for physics in the 1901's. Henri Becquerel, that came to know of this discovery in a seminar at the Academy of Sciences, was so fascinated to devote himself to the study of this absolute novelty.

At that time, while studying phosphorescence, which is the emission of natural light of some elements after exposed to sunlight, he supposed that this property had some commons with X-rays.

He received a gift from his father: some uranium salts that became phosphorescent when exposed to sunlight. He randomly arranged them, before exposing to sunlight, near a photographic plate to be used to test the emission of X-rays and, unexpectedly, on the plate appeared the shadows of uranium crystals: the plate was impressed without the salts having been previously exposed to sunlight. Subsequently, by interposing other objects of various forms and materials, between the plate and the crystals, he always checked the appearance of the shape of the object on the plate.

This meant that the image could only come from the irradiation of the uranium salts, regardless of the absorption of the solar rays and therefore, the uranium salts emitted radiation independently without any external stress. All this led to the conclusion that there are elements that spontaneously emit radiation.

In 1899 and 1890, starting from this random event, Becquerel studied the radiation emitted by these elements and verified that, unlike X-rays, the radiation emitted by uranium was made up of charged particles because they suffered a deviation, if affected by an electric or magnetic field,.

He saw that this radiation was decreasing in intensity over time. Now sure it was a new discovery, he called them *Becquerel rays*, but later changed in "radioactivity".

In 1902's Rutherford and Soddy explained radioactivity as the spontaneous process of atomic transformation.

At that time, a young student, Marie Curie, was looking for a topic for her doctoral thesis.

Marie Curie, whose maiden name was Maria Salomea Skladowska (Warsaw, 7November 1867-Passy, 4July 1934), was physics, chemistry and mathematics of Polish origin later nationalised in France. She was the daughter of a professor in physics and mathematics and a teacher and musician. She moved to Paris from Poland to deepen her scientific studies. She had studied mathematics and physics and planned to return to Poland to start working as a teacher. To pay for his studies in physics, mathematics and chemistry at the University of Paris, she took evening classes and starved, literally. She earned so little money that ate bread, butter and tea. She met Pierre Curie while looking for a laboratory for his experiments and was introduced to him by one of his professors. Pierre, although did not have much space, agreed to share it with her.

Pierre Curie, enthusiastic man of science, changed Marie's plans that decided to marry him and pursue his career with a doctorate. When they were married in July 1895, Marie wedding dress was a dark blue dress that later she used as a lab coat.

Maria Curie, was so interested in the studies and results of Henri Becquerel that choose them as the subject of her doctoral thesis.

During the experiments she used the electrometer, a tool designed by his husband Pierre, which measured weak electrical currents.

Having acquired a small amount of *pitchblende*, a crude mineral made up of uranium and thorium, another radiation emitting element, extracted in the present Czech Republic, she verified that it was much more radioactive than the amount of uranium and thorium it contained: there was something else that was leaking. In July 1898 they received other confirmations with the discovery of a substance 300-times more active than uranium, which they called polonium in honour of the nation of origin of the scientist.

Marie Curie called *radioactivity* this ability to emit spontaneous radiation from certain natural elements such as uranium, thorium, radio, etc.

In the 1903 Pierre and Marie Curie shared with Becquerel the Nobel Prize awarded as recognition for *the extraordinary services rendered by his discovery of natural radioactivity* and, in the 1911, Curie again obtained for the discovery of radio and polonium.

In the 1925 Irene, daughter of Marie and Pierre, joined her mother in the radioactivity studies, and in the 1928, along with her husband Frederick, she experimentally identified both positron and neutron. Due to the lack of interpretation of the findings, this was later attributed to Carl Anderson and James Chadwick.

In 1935, again with her husband Fréderic, Irene Joliot-Curie received the Nobel Prize in Chemistry thanks to the discovery of *artificial* radioactivity.

Marie Curie was the first scientist to experience radioactivity and its health hazards that were unknown at the time, so much that it was considered a solution for many illnesses and was used in antirheumatic ointment, soaps, toothpastes, etc.

The radioactive test tubes were transported freely without protection and were often placed in the pocket and drawers of the desk. In addition, during the Great War, the scientist provided assistance as a radiologist in hospitals and, even there, absorbed large quantities of X-rays, always working without protection. And so Marie Curie died on the 4th of July, 1934 from aplastic anaemia caused by radiation absorbed in the experiments.

In the course of her life Marie experienced triumphs but also great humiliations. In the 1903, despite the achievement of the Nobel Prize and the recommendations of many of her physical colleagues, the French Academy of Sciences denied her the honour of entering the staff to defend the nationalism, because of Polish origins, because of sexism as a successful woman and for anti-Semitism also, because her maiden name could be considered Jewish in origin.

In 1995 her human remains and those of his husband were moved from the cemetery of Sceaux, south of Paris, to the

Pantheon of Men and Illustrious Women of Paris. Mitterrand. The French President at the time, in his speech remarked how this woman entered the Pantheon for "her merits", as the first doctor in science, the first professor of the Sorbonne and also the first person to receive two Nobel prizes.

Marie Curie, the woman who fought the prejudice of time by fighting depression, was an intellectually severe and perfectionist, with her work apron on her black dress, just like so many of our grandmothers, was like a child stunned in front of science.

> *I am one of those who think that science has great beauty in it.*
>
> *A scientist in his laboratory is not only a technician: he is also a child placed in the face of natural phenomena, which impress him as in a fairy tale*
>
> *Marie Curie*

The studies of both Curies and Becquerel launched important news: there were atoms that emitted charged particles and these were present in all the atoms of all elements.

At the dawn of the new century, the twentieth, was reached the knowledge that there were stable elements, composed of atoms that behaved in a neutral way despite having negative particles and that therefore should also have positive particles to balance the charge and, there they were elements composed of atoms that spontaneously emitted radiation.

The scientific community left the nineteenth century ignoring how the positive and negative charges were arranged inside the atoms.

The Big Race

> *The universe is full of magical things patiently waiting for our wits to grow sharper.*
>
> *Eden Phillpotts*

1900 The Universal Exposition in Paris

The 20th century presents it to the starting blocks ready to achieve great goals and Physic Science is certainly one of the main competitors from who are expected the greatest results. Both the industrial modernisation and the new social balances make us forget the long period that has passed without great scientific results. In the air you experience the symptoms of a change that will become history.

Throughout the world there were at most a thousand physicists who participated actively in public life: they were representatives of a class of the bourgeoisie that was usually well paid and that was easy to find in society meetings or in the cultural lounges of the various exponents of the aristocracy. With their presence they guaranteed prestige and honour and many were fascinated by the stories of their experiments and new discoveries. Napoleon III, William II and the King of England, only to mention the most famous, often received the most illustrious scientists to make them participate in their discoveries.

Europe, in that historical time, was a pioneer in all fields of science, represented the best of scientific innovation and England, Germany and France were the elite. The Asian powers we know today do not play a significantly appreciable role: Japan was gradually emerging from its long Middle Ages; French pride, after the humiliating defeat of Sedan of the 1870's, found the right incentives by focusing on scientific innovation; Germany's industrial power, supported by an ever-increasing military power, rivaled on a par with that of England.

The historical period that began from the 1870 led to the second industrial revolution. The capitalist economy transformed irreversibly the society of the time that saw the spread of many new artefacts made available to an increasing number of people. All these increased product demands required the industry to modernise with new production techniques and machinery. The ever-increased distribution of products and production changed the balance of power between the manufacturing sector and the centralised state powers. Above all, there was a need to review international economic relations, based on the new maps of the world industrial establishment.

The policy of free competition, thanks to an ever widening international market, of the new motorised means of transporting goods by sea and land along the new routes of connection of the Suez Canal and the Panama Canal, became anachronistic. In order to meet the challenge of the larger and faster demands of the ever-expanding, but very distant, new markets, more and more investments were needed and that was impossible for small entrepreneurial realities. So the entrepreneurs sought new solutions.

So grow up the first holding companies and the first business agreements that influenced production and prices and, of course, the first trusts between previously independent companies. In some cases were created real monopoly regimes, especially in the United States and Germany, which were evolving into global economic and industrial powers.

Since 1879 Germany's rampant protectionism has forced Britain's economic policy to undergo major changes in the search for new markets, especially overseas.

The progresses were also changing gradually the way in which the science was felt. Small and large realities experimented and proposed new theories. There was a time when genius could still do without sophisticated instrumentation. The new path taken by science needed more and more electricity, but in the 1895 there were very few experimental laboratories that had a constant supply of electricity and so they helped themselves with accumulators and batteries. The more qualified a laboratory was, the more it was able to store electricity.

Thermodynamics, a branch of physics, was considered very well established thanks to well documented empirical laws and axiomatic bases and for the most scientists, it had nothing to do with atoms or statistics.

The publications produced by the physicists of the time mainly covered liquefaction of gases, optics, sometimes electromagnetism and, recently, electrical discharges into gas and vacuum tubes. Gibbs and Boltzmann's kinetic gas theory was a field of development for very few believers.

The inventions and discoveries of this period later marked the path of our history

This period is followed by inventions and fields of industry that will develop large gains:

Inventions	Field of Applications
Internal Combustion Engine	Iron and steel industry (cheap steel obtained from the residues of bauxite processing)
Light Bulb	Electrical Industries
Car Tires	Chemicals (new manufacturing processes that allowed the mass production of cheap chemicals)
TNT	Esplosive (dinamite)
Acetylsalicylic Acid-Aspirin	Drugs (Bayer)
Plastic	Petrochemicals

Electric elevator, telephone and gramophone, typewriter and bicycle, electric tram and many other smaller household items are also instruments that were invented at that time: when we use them daily we should remember how recently they have been discovered.

The second industrial revolution had been less dramatic than the first, but more widespread and capillary and, irreversibly, changed habits and behaviours, simultaneously initiating that process that took the name of consumerism. These changes contributed to a more open mind and intellectual contempt than ever before. Scientists who experienced these changes were also influenced by them and began to think of new theories and extraordinary experiments.

Humanity was ready to fly higher.

The real innovation of this period is the increasingly strong link between science, technology and industrial production. The new patrons were not only the rich aristocrats but universities, industrial research laboratories and especially the military.

All these motivations explain both why a new class of minds was born that was interested in physics, and why it changed the way of doing research: no longer "pure" and unconditional, but aimed for the purposes of third parties, almost "dirty" by other interests. It was the first warning of a S*ystem Physics*.

Or more generally, a *System's Science*.

Most of these changes took place in the old Europe, perhaps because of a middle class bourgeoisie which has its roots into the enlightenment and in the changes it produced. This could also highlight how, even unconsciously, all scientists of any subsequent period have been, more or less consciously, "directed" or "conditioned" to achieve pre-ordained goals. Having to respond of the results and objectives prevented free research and even more at present times. This can only get worse, although it seems that the results obtained are always high-level and free from external conditioning.

1904 The Plum Pudding Atom Model

The beginning of the 20th century was expected with hope and optimism. The Universal Exposition of Paris, from April 14th to November 12th, was expected to be of great hope. Participating countries exhibited the best results, especially in the field of technology, with the most recent inventions.

Walking through the great Pavilions it was easy to meet famous characters of that period and, probably, the two brothers Auguste and Louis Lumière stopped to talk with the charismatic Jules Verne.

During the opening period more than fifty million persons visited it. During September, 1901 Guglielmo Marconi, that already invented the wireless telegraph, tried and realized the impossible first transatlantic radio transmission. On this occasion, too, there was an obvious case of serendipity: luck played a decisive role.

Instead to achieve the wrong predictions that radio waves follow the Earth's curvature, the experiment succeeded because the waves were reflected by the ionosphere, as Appleton later confirmed and, reached the receiving station, involuntarily decreeing the success of the initiative.

The telecommunications industry was originated from this event.

On the 17th of December 1903, the "Flyer", of the American brothers Wilbur (1867-1912) and Orville (1871-1948) Wright, the first flight took off, travelling for 36,5 meters in twelve seconds an reached three meters from the ground.

This was the initially step for the aviation industry.

At that time physicists wondered about the conclusions reached by Thomson that in the 1904, strong of the experimental results, proposed to the scientific community the first atomic model.

His theory was opposed with great scepticism and was impeded almost universally.

This atomic model, later known as the *Pudding model*, supported that the atom had a series of characteristics: it had a total neutral charge, it was similar to a sphere, or preferably to a cloud, it had a positive charge with embedded, negative charge particles, later identified as electrons, just as if they were raisins in a pudding.

This model adapted to many of the experimental evidences of the work carried out by physicists and chemists up to that time, but was unable to respond to the most recent experimental observations that appeared to involve light emissions from some atoms.

Albert Einstein was about to arrive.

But Thomson was quickly overpassed from the *Rutherford model*, one of his students at the *Cavendish Laboratories*, because discovered the atomic nucleus starting the experiments on particle deflection α.

The concept of electron will undergo considerable changes in the years to come, but it will always remain a fundamental part of the atom. From the year of its discovery, in 1897, began the analytical study of the elementary constituents of matter.

A curious story: in the 1906, the Nobel Prize was awarded to J. J. Thomson for the discovery of the corpuscular nature of electron; in 1937, his son G. P. Thomson, on the other hand, won the Nobel Prize for highlighting the wave nature of electron! These children who always contradict their parents...

Thomson's son, with his theories, made a considerable contribution in the development of the technical bases for the realization of what will become, thanks to Aston and others, the *mass spectrograph* and hypothesized the existence of **isotopes** of atoms which he succeeded in confirming experimentally.

In those early years of twentieth century, thanks to men with great imagination and without fear of what we would today call "media pillory", new ideas, hypotheses and theories emerged so innovative and in such a rapid succession that it was almost impossible to reconstruct the temporal flow of the succession of events. It is still difficult for me to understand how, from the 1900s to the 1925s, such an important number of interpreters of physical theories could be concentrated, not only intelligent but brilliant.

Experiments have led to the formulation of theories that, starting from the numerous experimental results, tried to explain the knowledge of the Universe in a completely different way. What is most striking, in retrospect, is the flexibility with which old ways of thinking were abandoned in order to deal with the solution of questions, until then completely ignored, with new interpretations explaining the surrounding physical world.

How was all this possible? Many have succeeded not only in improving themselves but, literally, in abstracting themselves with incredible theories.

The *humanity* rised from experimental evidence to the abstraction of theoretical models, first intuited, then predicted and finally verified and tested.

The physics that was developed since then maintained expectations and has become what we know today because there were these hypotheses and theories or, if there had not been, would we have achieved different results, with other protagonists, perhaps later but perhaps more adherent to nature?

Were they the right men, or were they the cause of a change that we do not yet understand today?

At first glance it would seem that instead of explaining nature, working with nature, we went through it, raping it in its essence.

These pioneers have always risked themselves by putting their faces to what they said. Their ideas have spread to today's physics: but were these the most correct ideas or were they the ones that were most convenient to answer?

"Quantum Physics"

> *All matter originates and exists only by virtue of a force that makes the particles of an atom vibrate and that holds together the tiny solar system of the atom ... We must assume the existence of a conscious and intelligent mind behind this force. This mind is the matrix of all matter.*
>
> *Max Planck*

Planck: Quantum Physic's Dad

Richard Feynman argued that

> *... there are two kinds of geniuses: the 'ordinary' and the 'magicians'. An ordinary genius is a fellow that you and I would be just as good as, if we were only many times better. There is no mystery as to how his mind works. Once we understand what he has done, we feel certain that we, too, could have done it. It is different with magicians. [...] the working of their minds is for all intents and purposes incomprehensible.*
>
> *They seldom, if ever, have students because they cannot be emulated and it must be terribly frustrating for a brilliant young mind to cope with the mysterious ways in which the magician's mindworks.*

Karl Ernst Ludwig Max Planck (Kiel, 1858 - Gottinga, 1947) became father by the will of physicists after his time. His assumptions upset an old status quo that many colleagues had tried to explain without success. His preparation and vision of classical physics, coupled with a careful study of the results of experimental physics, enabled him to understand and explain a series of results distancing him from the ideas of his contemporaries and led him to propose theories that would become the basis of a whole new physics that, from that moment and for a long time autonomously, it would develop in parallel with classic physics.

All objects present in nature absorb and reflect light from the surrounding environment and a black body reflects minimal amounts of light and tends to warm up faster and much more than other bodies.

Normally we can verify that by heating any object, after a while, it becomes incandescent and thus emits light energy. In fact the object begins to emit radiation even before there is a visible effect, radiating into the infrared spectrum that we cannot see with our own eyes. The reason for this is that, while the body warms up by absorbing light energy, the electrons on the surface of the material, agitated thermally, are excited and disemboweled by radiating light.

At the end of the 19th century scientists began to examine how the rays of a black body behave according to its temperature.

To facilitate the calculations it was postulated a perfect black body, which is an ideal object that does not exist in nature and that appears black because it absorbs all the light it receives. Heating a black body will irradiate it by emitting light.

Since there is nothing truly black, the scientists of the time imagined and constructed an empty cavity with a hole from which to put the light inside it. The light entering was reflected from the walls and absorbed, except for a negligible amount that would have escaped from the same entrance hole.

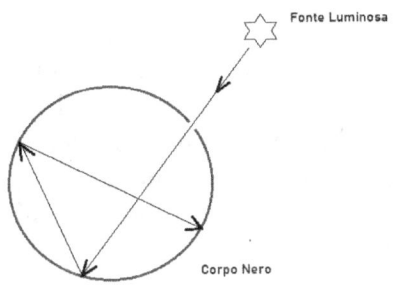

The empty cavity absorbing the input energy strive to heat up more and more and at the same time emit electromagnetic radiation.

The results of this irradiation said that it was possible to establish a theory that would predict its course. But all physicists were in great difficulty, because it was not possible to find a single law covering the results of all measurements made. The classical method used up to then only explained the low frequency data.

In order to arrive at a unified solution, therefore, there was a need to completely change approach and point of view.

Planck's great merit was to change the interpretative pattern of high frequencies without any justification, either mathematical or scientific, but only on the basis of an intuition, with the courage to formulate a postulate that explained all possible energies as multiple of the value of a single base energy.

This is a fundamental step, because the method of study is radically changing: there was a problem that had partial solutions. All of his colleagues focused on the possible solution using only mathematical instruments and physical laws available, needlessly, without exploring other possibilities.

At this point the scientist has only two choices: either the problem is insoluble, but this is not part of the general scheme of nature where everything seems to obey a law that can be hidden but that exists, or one must think in an unconventional way, risk being ridiculed and propose a solution that includes all the data but may also not be shared.

A general law that should always be present in a researcher's room is the following:

> *In physics, if there is an unsolvable problem for everyone, there will certainly be a man who will one day solve it.*

When this happens, we are no longer only at the presence of a scientist, but also of a genius: a normal person, until that time that sees a different solution following paths that do not exist for others.

This happened for Planck, that relating energy and frequency, imagined that the energy of the electromagnetic wave was emitted not randomly, but only as multiples of an elementary quantity. The multiplier was a very small, but well identified number, which from that time was called Planck Constant. It was just a mathematical artifact that allowed an explanation to be found, but it worked every time this method was applied.

This was the initial intuition from which "Quantum Physics" developed and since then the physical world has never been the same: and consequently the description of microscopic phenomena changed radically.

Let us imagine what went through Planck's mind: solving the problem of the radiation of the black body, the physicist immediately realized that small shift to the amount of energy would have consequences in any theory from that moment on! Not only that, but probably many previous conclusions should have been revised in the light of the discovery of that small multiplier.

Everyone had to shake his wrists and in fact, he did not immediately publish this intuition fearing the judgment of his colleagues.

On the December 14, 1900 Planck, winning all fears, decided to make the scientific community participate in its conclusions and published them.

That was the official date of birth of "Quantum Mechanics" and Modern Physics.

In his memoirs he confirmed all his doubts and fears referring to the development of his theory on black body radiation, illogical to the knowledge of the period and, he wrote:

> *The whole affair was an act of desperation...*

> *I am a quiet man, naturally opposed to rather risky adventures. But ... a theoretical explanation had to be given, whatever the price ... In the theory of heat it seemed that the only things to save were the two fundamental principles (conservation of energy and principle of entropy), for the rest I was ready to sacrifice all my previous convictions.*

For this contribution, Planck was awarded with the Nobel Prize in 1918.

There was another aspect of his method, which he himself highlighted in his memoirs and which should be adequately re-evaluated by all scholars who want to propose new theories and not just explain the previous ones: it was the first time that the inconsistencies, due to application of obsolete laws of physics, were circumvented thanks to

> *A fortunate purely mathematical violence against the laws of classical physics.*

Today everyone likes to say how solid science is Quantum Physics and how it produces very good results: but how many people know this so artificial starting point?

Can we consider an example of serendipity also this or did the mathematical result of the experimental evidence bend to the will of the scientist?

Maybe "Quantum Physics" wanted to give the first sign of its potential. For the first time, it was not empirical evidence that explained a physical phenomenon, but it was the physical phenomenon that adapted to the mathematical formula. I believe that in addition to quantifying energy, we can also think to quantify the understanding of physics.

At this stage, questions are not really rhetorical and we must thank Planck's "irrationality" which, finding the constant mathematics that bears his name, has allowed a solution to a problem that until then was insoluble.

But what would have happened if there had not been a Mr Planck and then a Mr Einstein or any other genius who had not risked before or after?

The question again arises: which direction would physics have taken?

Would "Quantum Physics" have been an unquestionable development anyway, or would it not have existed at all?

Is this logical artifact perhaps the cause of a blind alley in which modern physics seems to have gotten lost and which has prevented perhaps more efficient theories and developments?

Is it a coincidence that from that moment on, and from that intuition, physics seems to receive an impulse that leads to an obvious acceleration?

The question that haunts more or less consciously so many scientists is that, if everything seems to become simpler, could there be a more hidden, more rigorous path that we have not taken into consideration and that is regardless of an explanation based on an artifact, as recognised by its own discoverer?.

The simplest answer is that the artifact was the stroke of genius and that all the follow-up is the consequence of that. And everyone will remain irrevocably in their own positions.

However, the consequences were obvious from the start. From that moment on, mathematics changed its role. New formalisms were needed to better integrate the numerous results of the experiments that were designed and developed. Mathematics had become an instrument of physics, just as a pawn for a peasant is.

Planck died almost ninety-nine years in Gottingen, and in his last conference a few months before to die, he made his will for anyone who wishes to become a scientist:

> *Those persons involved in the construction of the sciences will find their joy and happiness in having investigated the questionable and honored the unobservable.*

How useful it would be for our contemporary physicists, who insist on pursuing results only by relying on old ways of working and are convinced that they are only using comfortable paths already trodden, to go and reread a bit of the history of physics. Surely they could rediscover the spirit that should be behind every physicist and maybe, a scholar could transform and finally become a genius.

Scientific World Reaction's

Albert Einstein (March 14, 1879 - April 18, 1955) was a German physicist. In those times he was a young man twenty one years old. Some years later, his words will make us understand the amazement and the extent of the discovery of Planck's Constant:

> *It was as if we suddenly lacked the ground beneath our feet and saw nowhere solid ground on which the foundations of a new construction could be laid.*

Initially Planck's theory was not well received by the scientific community, because it seemed to be more of a hypothesis derived only from a specific experiment, suitable to explain the results of one very particular phenomenon, but which had no other possible application. Planck himself, at least initially, regarded the result as a limited explanation for the radiation of the black body. It was Albert Einstein in 1905 which formalized the universal goodness of the new "Quantum Theory" applying it to the interpretation of the photoelectric effect. In this way a mathematical artifact took charge of a theory and officially became a physical greatness available for all.

When Niels Bohr used it to confirm his atomic model, it became fundamental to the development of modern physics.

Max Planck's result changed the way to understand physics and led to a change in the studies of nature: from then the phenomenology of natural events, especially in the microcosm but, as will be discovered much later also in the macrocosm, will be treated and interpreted in accordance with models based on *discontinuity*.

It is curious to know that Planck, despite his genius, his affirmations and his achievements, never had follow-up and the students who chose to study with him were always few. Perhaps because Planck himself did not immediately believed in his theory, so much so that he came to affirm the despair of having to invent this artifact?

Why did Planck say that?

Because he had been forced into a revolutionary act. This act contradicted strongly the classic laws of physics: the quantum model predicted that light would be absorbed and emitted by matter no longer continuously, as wave theory wants, but discreetly. Packaged.

These multiple energy units of a basic value, the, were called *"Quanta of Energy"*.

Substantially his contemporary scientific world preferred to ignore the discovery, considering it an illusion to be used by a theorist who would soon portray it. Excepted a young boy working at the Office Patents in Berne, passionate reader of all new articles on physics, He was fascinated by the plant put up by Planck and his innovative hypotheses. He was Albert Einstein.

For about five years Einstein continued to study almost as a self-taught and, as unknown person for most of the scientific world, published an article on Annalen der Phisik that proposed an explanation of photoelectric effect, in which justified the electrons that were released by the metal excited by light radiation, using Planck's quantum numbers for understanding.

Einstein also used Planck's hypothesis to affirm that light not only interacted with metal through energy *quanta's* but itself was made up of these discrete packets which from 1926 were called *photons*[5].

With the experimental confirmations of his theory, Planck reassured himself: he had been right, imagining something incredible, where others had not been able to observe and understand.

In 1905 Einstein lived his *Annus Mirabilis*, and later received a Nobel Prize for this discovery.

The academic and scientific world was forced to change its mind and from that moment the research on quantum and *"Quantum Physics"* progressed to revolutionize and replace in a few decades classical physics, which cannot be used to explain the atomic world.

The result achieved by Einstein for many would have been a point of arrival, instead the scientist continued to study, coming to propose an even more advanced theory that predicted the behaviour of electromagnetic radiation in the cosmos; and thus also gave impetus to research on gravitation.

However, doubts about the behaviour of subatomic particles continued to remain partially unresolved: the duality of their behaviours was irreconcilable with the results of some experiments and left unresolved the question of whether they were corpuscles or waves. Logic wanted them to be particles, but

---

[5] The *photon*, from the Greek φῶς "phòs", light, was used for the first time in July 1926 by the physicist Frithiof Wolfers. Subsequently the chemist Gilbert Lewis reused it making it definitively enter the physical terminology.

the light emission of some elements hinted at wave behaviours. If this were true, was light a component of matter, or vice versa?

Or was there something else that no one imagined and the scientists were making a mistake?

Investigations on the Nature of Light

Light, from original Latin word *lux*, is the visible portion that human eye can see of the electromagnetic *spectrum*, which is energy that travels through waves. The entire *spectrum* is very large, and it goes from very long *radio waves* to very short *gamma rays* and includes all types of light, even those that the human eye cannot see, because only a small range of this spectrum, called visible light, can be seen. The limits of the spectrum visible to the human eye can vary from person to person. All wavelengths simultaneously visible merge into a single luminous beam called *white light*.

Man has always assumed that one of the fundamental properties of light is to propagate in a straight line. Many optical phenomena can be described by the laws of reflection and refraction. Instead, when light interacts with live edges, edges or narrow cracks, it is necessary to refer to *interference and diffraction phenomena*.

The *diffraction* was described for the first time in the 1665 by Jesuit astronomer Francesco Maria Grimaldi, after he had observed that not all optical phenomena were describable with the only laws of *reflection* and *refraction*.

Isaac Newton and Christiaan Huygens were great antagonists in their time and sparked a heated debate. Newton believed that light was composed of small corpuscles propagating in a straight line; Huygens, on the other hand, claimed that light was similar to sound waves and wanted to interpret the results of the observations as if they were due to waves.

Newton's Corpuscular Theory considered that light was composed of particles emitted from a light source in all directions. The laws of this movement belonged to *classical mechanics*: the propagation was in a straight line and the corpuscles bounced against each other or against obstacles, causing the phenomena of reflection and refraction.

This model could explain the formation of shadows, eclipses and also rainbows, which was due to so many coloured strips due to the decomposition through a medium that acted both as a refractor and as a reflector of the particles. The white light was the compaction. Newton was able to prove this idea by inventing *Newton's Disk*, a circle divided into seven parts with the colors of the rainbow: red, orange, yellow, green, blue e, purple and indigo. By spinning the disk, the reflected light recomposes itself in an almost natural coloration, thus giving the observer an illusion of white.

Christiaan Huygens, who instead favoured Corpuscular Theory, saw the light beam as a wave in propagation through a medium that associated with the already existing concept of *Aether*. According to his idea, this element permeated the entire Universe and consisted of an *enormous amount of elastic particles* that facilitated the propagation of the wave. The speed of light changed because the wave, crossing media with different densities, was influenced by it. This theory was also applicable to almost all phenomena of light, including reflection and refraction, but was generally more complex than Newtonian Corpuscular Theory.

It was only in 1864 that James Clerk Maxwell (Edinburgh, 13th June 1831-Cambridge, 5th November 1879), a Scottish physicist, with his publication *A Dynamical Theory of the Electromagnetic Field*, assumed that all previous observations, experiments and equations on electric, magnetic and light fields, could be correlated in a single theory that we know as *Maxwell Equations*. This was the first modern T*heory on Electromagnetism*.

Maxwell had appreciated the 1820's experiments in which Hans Christian Oersted hoped to find a connection between magnetism and electricity, especially after the experiences of Michael Faraday that in 1831 had succeeded in converting electricity into magnetic energy using an isolated wire and a galvanometer. Maxwell, starting from Faraday's conclusions, realized that an electromagnetic wave was traveling at the speed of light. Thus he was able to incorporate light, magnetism and electricity into a single theory.

Despite all evidences and all demonstrations that tended to regard light as a wave phenomenon, without considering the experimental tests provided as definitive, scientists preferred Newton's mechanical-atomic theory until the end of the 19th century because it was the most consolidated theory. And also it best suited to both the general physical ideas of that time and the calculation habits, according that light was a set of particles that obeyed the law of gravity.

Physics, once again, could not shake off the habit of adapting to working theory, rather than face the risk of seeking the best but not yet consolidated theory.

Young's experiment: the most beautiful

It was Thomas Young (Milverton, 13th June, 1773 - London, 10th May, 1829), a British physician that convincingly demonstrated the wavy nature of light. The chronicles of the time spoke of him as a child prodigy and later as an eclectic enthusiast of every branch of knowledge. At the age of twenty-eight, he left the medical profession and joined the Royal Institution of London. Great linguist, he was among the first to decipher Egyptian hieroglyphics and contributed greatly to the decoding of Rosetta stone. Young distinguished himself in all the fields he applied: in physics he contributed to the wave theory of light, in engineering he proposed the module of elasticity, in physiology he explained the mechanism of vision, in linguistics he helped to understand the hieroglyphics and as linguist he discovered the common parts of the Indo-European languages. Young is considered and remembered as *"the last of the men which knew everything"*.

Unlike Newton, Young was convinced that light was a wave, because it suffers the phenomenon of typical wave interference.

From 1801 to 1803, in a series of conferences to the Royal Society, Young presented the idea of the principle of interference and experimental demonstration in support of wave theory. The experiment that proved his theory was considered one of the most "beautiful" of physics, because he managed to combine a wonderful practical simplicity with the demonstration of an extremely complex result: the first convincing demonstration of the wavy nature of light.

The necessary equipment consists of a light source, a sheet with two adjacent slits and a wall preferably dark.

We direct the beam towards the wall and illuminate it; now between the light source and the wall we put the paper with the two slits. The waves that emerge from slits, projected on the wall, overlap and on the wall appear a pattern of light and dark regularly spaced lines that constitutes the pattern of interference.

This was the first clear proof that: light superimposed by light can generate darkness.

The experiment confirmed Huygens' intuitive ideas about the wave nature of light: clear areas are constructive interference, dark areas destructive interference.

After almost a decade, the theoretical experimental work of A. J. Fresnel confirmed Young's experiment and showed that there were also many other effects that could be interpreted only from light wave theory.

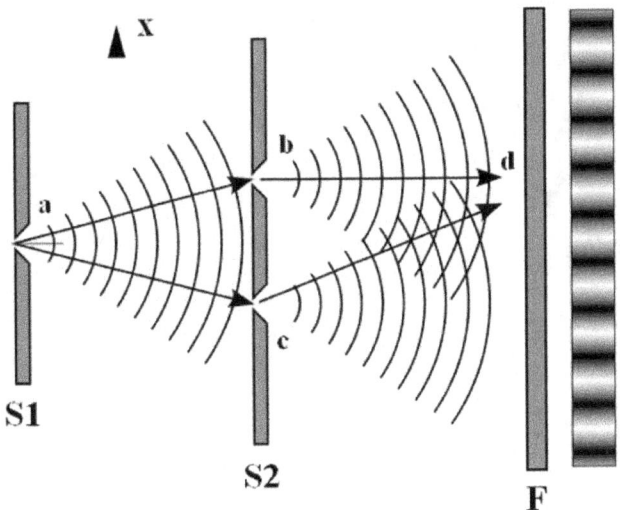

Double-slit experiment showing the interference of light. Source: Wikipedia

Einstein Grandmother

> *When I ask myself why it was that just I elaborated the relativity theory, the answer seems to be linked to this particular circumstance: a normal adult does not care about the problems of space time, all the possible considerations on the question have already been made in the early childhood, in his opinion. I, on the other hand, developed so slowly that I began to question space and time only after I grew up and consequently studied the problem more thoroughly than a normal child would.*
>
> *Albert Einstein*

Introduction to Relativity

Classical mechanics had linear logic patterns that were applied to the understanding of natural phenomena: what was seen was explainable with deterministic formulas that gave a certain result, as a consequence of the observations. There could have been different interpretations of a phenomenon, but the result was certain. The 20th century changes this pattern because it reveals unexplained phenomena or, even stranger, without any apparent connection to physics. The scientists contribute to this confusion by differentiating the methods of research of the same phenomena and often explaining them in contradictory ways, thus preventing a common narrative in the name of a single physical reality. One example of all was the difficulty of proposing a single theory on light waves that would make everyone agree.

Because of this, two branches of physical research began to coexist that still cannot converge today: the theory of relativity that modernized the concepts of reference systems and the interactions with electromagnetism and the new "Quantum Physics", which focused mainly on the interactions between matter and radiation.

Fortunately classical physics continues to be a valid solution to those questions posed by the observation of natural phenomena in daily experience, but the world of the extremely small, made of atoms and particles and, phenomena on an astronomical scale, present major problems of coexistence. Although physical realities are common to our Universe, it seems that they are not traceable to shared laws and so, even if they are sad, they take separate paths.

Among the first to break up this relationship was Maxwell with his theories on the electromagnetic field, which were not in accord with Newton's law of universal gravity, which tied distance and matter exclusively to the force of attraction directly proportional to the quantity of matter and inversely proportional to the square of distance.

The coup de grace came mainly from a young Albert Einstein, a stranger to the academic world because of his profound lack of authority expressed by his teachers and hierarchies in general. For these reasons he lost the opportunity to work as a laboratory assistant, but thanks to the father of a friend of his, he was hired at the Swiss Patent Office in Berne: an office job that gave him the free time needed to concentrate on his own research. Thus he managed to keep up to date on the progress of physics, occasionally discussing it with a close circle of friends and publishing occasional articles.

In 1905, his *Annus Mirabilis* published on Annalen der Physik, four astounding short-distance articles.

The Theories of Relativity changed forever the way of understanding space, time, mass, energy and gravity and, for the first time, it was noted that, thanks to Newton's laws, the motions of the stars and the gravity that governed them could be explained, without anyone ever having wondered what the cause of gravity was.

It may seem strange but Einstein's turn was first philosophical in the way he approached physical issues and then, only in a second moment, scientific path. He himself loved to point out, often saying that:

> *It is likely that without my philosophical studies I would not have come to the solution*

With this phrase he wanted to pay tribute to the philosopher David Hume, an empiricist and sceptic, author of 1738 *Treatment of Nature* who had contributed to the modelling of his forma mentis. Einstein had absorbed the convictions that would lead him to conceive new ideas about space and time. Other elements that Hume instilled in young Einstein were the convictions that beyond theoretical abstraction there must be certain experimental proof and that time does not exist unless linked to the movement of objects.

Einstein was now perceived as a fascinating meme: his physical presence was no longer necessary. He himself had become his ideas.

Even today his thoughts and physical theories, in the original formulation, are so far removed from our daily sphere that they are more reason for reflection than for study.

Moreover, it is also closer to us, common men on the street, because we instinctively perceived it, unlike other genes, as a character very similar to each of us, also because, despite the exceptional results of it, in the end it told anecdotes, ate bread and paid the newspapers like all us. He was, exceptionally, a common man.

To build his theories Einstein used tools at everyone's disposal. He studied what we studied without having any computers at his disposal: he only used pen, paper and brain.

The difference is all in his mental constructs and in his speculative methods that led him to achieve the known results differentiating him from all of us.

Being the details of Einstein's theories readily available, we will instead devote ourselves to investigating Einstein's method of working to find out, possibly, what were the mechanisms that activated in his mind to push him to reach these results?

Jacques Solomon Hadamard (Versailles, 8th December, 1865 - Paris, 17th October, 1963) a French mathematician, known for demonstrating the theory of the first numbers, was contemporary with Einstein. While preparing a psychological investigation into the mental processes of mathematicians, he also asked Einstein questions to try to understand his thought processes.

I bring back the original answer Einstein gave so that those who desire it can deepen their logical-mental mechanisms...

(A) The words or the language, as they are written or spoken, do not seem to play any role in my mechanism of thought. The psychical entities which seem to serve as elements in thought are certain signs and more or less clear images which can be "voluntarily" reproduced and combined. There is, of course, a certain connection between those elements and relevant logical concepts. It is also clear that the desire to arrive finally at logically connected concepts is the emotional basis of this rather vague play with the above-mentioned elements. But taken from a psychological viewpoint, this combinatory play seems to be the essential feature in productive thought--before there is any connection with logical construction in words or other kinds of signs which can be

communicated to others.

(B) The above-mentioned elements are, in my case, of visual and some of muscular type. Conventional words or other signs have to be sought for laboriously only in a secondary stage, when the mentioned associative play is sufficiently established and can be reproduced at will.

(C) According to what has been said, the play with the mentioned elements is aimed to be analogous to certain logical connections one is searching for.

(D) Visual and motor. In a stage when words intervene at all, they are, in my case, purely auditive, but they interfere only in a secondary stage, as already mentioned.

(E) It seems to me that what you call full consciousness is a limit case which can never be fully accomplished. This seems to me connected with the fact called the narrowness of consciousness (Enge des Bewusstseins)"

From "A Mathematician's Mind, Testimonial for An Essay on the Psychology of Invention in the Mathematical Field by Jacques S. Hadamard, Princeton University Press, 1945." in Ideas and Opinions.

From Einstein's words we understand how he is not aware of what this maieutic mechanism sets up, exactly as it happens for all of us.

When Einstein translated the guiding ideas of his theories, these goals were the results of a momentary intuition or the result of a series of non-perceptible intuitions that suddenly manifested themselves on a conscious level after being elaborated on an unconscious level?

Or were these the result of the accumulation of a series of intuitions metabolized over time and which produced a common thread that appeared suddenly but which intrinsically connected all the hypotheses from start to finish?

For many, all three possibilities are present.

Special relativity

Special Relativity is Einstein's most well-known theory and was first exposed in 1905. Among other things it talks about lengths that can shorten, moving clocks that can slow down and about the light that is the fastest thing in the Universe and there can be nothing that can surpass it.

Einstein realized that in the results of physics of his time, there were two apparent contradictions.

The first was derived from the principle of Galilean relativity, so physical systems are not influenced by uniform movement. Simplifying this means that no mechanical experiment can tell if you are moving in a uniformly accelerated straight motion.

The second contradiction, the postulate of the invariance of the speed of light, was born directly from the conclusions of Maxwell and from the consideration that the speed of light was constant at approximately 300.000 Km/sec: did Galilean relativity also apply in electromagnetism?

Einstein understood that would be no contradiction if he had given up the hypothesis of simultaneity: thus, two events that occur simultaneously in an inertial reference system seem, on the contrary, timeless, to those who are in another inertial system in motion at different speeds.

Beyond this hypothetical obstacle, all the other were only mathematical processes and from this intuition, to arrive at the drafting of one of the greatest scientific articles of the 20th century, on the *Electrodynamics of Moving Bodies*, only five to six weeks passed.

The development of a theory that seems to be concentrated in the final moment evolves in a completely different way.

Einstein has always stated that it took more than seven years to solve and accept intellectually, as well as psychologically, the two apparent contradictions.

It was a work already started as a student, when he was fascinated by Maxwell's fantastic theory that seemed to be almost a theory altogether and that he was growing old simultaneously with electricity, magnetism and waves.

General relativity

Einstein presented the *Theory of General Relativity*, which addressed the questions of universal gravity on the 25th of November, 1915 at the Prussian Academy of Science.

For many people, it is still today the most beautiful theory ever exposed.

For us, men grown up hearing talk of time, gravity, galaxies, black holes, which we have however seen documentaries or generally movies where somebody talks about speed of light or time travel, the concepts and terms used by Einstein are now familiar, but for his contemporaries and colleagues, it must have been inconceivable. Having to take note of such a new and untenable theory at the time must have been more than traumatic.

From 1907, while still employed at the Berne Patent Office, a series of events that happened in everyday life led him to reflect that an observer in free fall no longer feels his own weight from which came what defined *"the happiest thought of my life"*: as happens in an elevator, if you reverse the direction of acceleration you can cancel the gravity by reversing it, because acceleration creates a gravitational field. This conclusion is its principle of equivalence.

In 1915, after systematically identifying all properties derived from that initial intuition, he devoted himself to their cataloguing in order to formulate a more general theory. Time measures, slow-motion clocks and the influence of gravitational fields on the path of light rays that are diverted in the presence of large gravitational fields were also included.

In the case of limited relativity there are many sources directly attributable to Einstein, such as his notebook or publications or autographed notes, which allow us to understand the progress of his logic.

The interpretation of intuitions was often contradictory as for any of us: back and forth to try to square a hypothesis, faster or slower if a formula flowed or was hostile.

Perhaps the only truly different method is that Einstein established the results of his final calculations in advance and only then sought the equations that produced them.

It would appear that his approach to the solution of physical problems was the observation, the intuition of what might be the point of arrival, the mathematical formalism to arrive at demonstrating the correctness of the process. This is an important step in the perspective, which should still be taken into account today by physicists, of the comparison with *"Quantum Physics"*: should the origin of a hypothesis be more mathematical or physical?

When Einstein discussed it, he argued that *"formal thought"*, as he himself defined the mathematical process, develops from equations that describe or could describe a theory, and thus hopes to quickly explain the physical system that is the object of observation.

The *ratio* of this method is the confidence that mathematics has an original intrinsic intelligence that simplifies physical problems. This logic sinks the roots into Plato for which existed a higher entity from which ideals and forms originate, eternal, incorruptible and immutable. From this ideal perfection derives the world that surrounds us, that at the moment it becomes perceptible loses its divinity by giving in to the corruption of nature until death.

If, on the other hand, we started from a physical hypothesis, we would have the certainty of the results of our experimental observations or, as in the case of relativity, of principles already widely demonstrated, such as the principle of relativity of energy conservation and quantity of motion.

Einstein preferred this second method, which already preceded the controversy with the School of Copenhagen: determinism against probability.

The Einstein method was to use mathematicians instrumentally, asking them to make tools, until then also non-existent, to achieve the idealized physical model.

Einstein did not come to his theories thanks to a mathematical model, as quantum physicists will, but he *"invented"* a physical hypothesis and asked mathematicians how to interpret it with functions that had as simple a formalism as possible.

This can be confirmed in the first publication of its first English translation of his *Über die spezielle und die allgemeine Relativitätstheorie,* On *the theory of special and general relativity*, in which on 1920 he states:

> *THE present book is intended, as far as possible, to give an exact insight into the theory of Relativity to those readers who, from a general scientific and philosophical point of view, are interested in the theory, but who are not conversant with the mathematical apparatus of theoretical physics. The work presumes a standard of education corresponding to that of a university*

matriculation examination, and, despite the shortness of the book, a fair amount of patience and force of will on the part of the reader. The author has spared himself no pains in his endeavour to present the main ideas in the simplest and most intelligible form, and on the whole, in the sequence and connection in which they actually originated. In the interest of clearness, it appeared to me inevitable that I should repeat myself frequently, without paying the slightest attention to the elegance of the presentation. I adhered scrupulously to the precept of that brilliant theoretical physicist, L. Boltzmann, according to whom matters of elegance ought to be left to the tailor and to the cobbler. I make no pretence of having withheld from the reader difficulties which are inherent to the subject.

On the other hand, I have purposely treated the empirical physical foundations of the theory in a "step-motherly" fashion, so that readers unfamiliar with physics may not feel like the wanderer who was unable to see the forest for trees. May the book bring some one a few happy hours of suggestive thought!

December, 1916   *A. Einstein*

Einstein considered his mathematical language understandable to a university freshman and, he addressed not only the scientific world but in the narrative part also to philosophers and enthusiasts. Not to mention that, by his own choice, he will often repeat the concepts to strengthen their understanding at the expense of elegance, which he leaves to cobblers and tailors.

Another consideration to keep in mind is that much of Einstein's work was done before his 30s. It would seem that the results obtained depended on the cunning and readiness of a young mind: if this consideration were true, we should better capitalize the potential of our young people.

Another element that emerges from studying his methods is the frequent use of mental experiments.

For Einstein

Imagination counts more than knowledge.

The Theory of Relativity must be addressed without dwelling on the mathematical aspects, but favouring the imagination guided by mental experiments.

Experiments, whether physical or mental, must above all induce us to

question ourselves without prejudice when a new experience conflicts with acquired knowledge

Among the notes he wrote in his own hand, I chose this short article that suggests, directly from the words used, his method:

The fact that I had neglected mathematics to some extent had its cause not only in my stronger interest in the natural sciences than in mathematics, but also in the following peculiar experience. I saw that mathematics was divided into numerous specialties, each of which could easily absorb the short life that was granted to us. As a result, I saw myself in the position of Buridan's donkey, which was unable to decide on a particular bundle of hay. Presumably this was because my intuition was not strong enough in the field of mathematics to clearly differentiate the fundamentally important, what is truly fundamental, from the rest of the more or less superfluous scholarship.

Furthermore, my interest in the study of nature was undoubtedly stronger; and it was not clear to me as a young student that access to a deeper knowledge of the fundamental principles of physics depended on the most intricate mathematical methods. This only came to my mind gradually after years of independent scientific work.

True, physics was also divided into separate fields, each of which was able to devour a short life of work without satisfying the hunger for deeper knowledge. Here, too, the mass of insufficiently linked experimental data has been overwhelming. In this field, however, I soon learned to smell what could lead to the fundamentals and to distance myself from everything else, from the multitude of things that clutter the mind and distract it from the essential.

Albert Einstein, Autobiographical Notes

Einstein formulated his theory from Newton who had come to the conclusion that two bodies attract each other because of a force called *gravity*. The mass and distance of each of the two bodies determines the intensity of the force.

The Earth attracts us by allowing us to stay with our feet firmly anchored to the ground, but also we, much weaker, attract the earth which, having an enormously higher mass, does not suffer from it. The same thing happens between celestial bodies: we can verify the effect of the Earth-Moon attraction in the tides.

From these evidences one understands how *gravity* is independent of the distance of the two bodies and, this is what happens for any cosmological element: all must undergo *gravity*.

Newton meant space and time absolute: any observer everywhere saw the same thing happening at the same time. Einstein's space and time were not absolute, but related to the observer's reference system and hence the name *relativity*.

Einstein in *special relativity* had already correlated space and time. General Relativity extends the understanding of cosmos by inserting gravity: large objects that are located in a Universe identified by space-time coordinates cause distortion in the continuum generating a force called *gravity*.

Extremely massive bodies, much more than our Sun, exert attractions so strong that depression degenerates into holes: black holes. These are innovative ideas very difficult to understand.

Einstein arrived at these results thanks to a combination of factors: certainly superior intelligence to the average, stubborn perseverance and also a capacity to manage all others who helped him to progress in the evolution of his thought, especially mathematicians.

Today, with current methods of study, it would be difficult to achieve or pursue Einstein's results: he was, at that particular historical moment, the right man, in the right place, at the right time.

He was himself, with his humanity and his ingenuity, the best example of relativity to a physicist waiting for a genius.

In the last twenty-five years of work Einstein did not produce such significant results: he sought in vain a unified theory that also included *"Quantum Mechanics"*, which he never accepted in the proposed formulation.

Perhaps his task had ended with the work he had presented and the transition to "Quantum Theory" implied another kind of thought that led to the passage of a witness to Niels Bohr.

Today Einstein is comparable to a legendary meme: in a very far future, someone will probably wonder if he really existed or if it was a way of justifying the magnitude of the intuition of *Relativity* by a scientific community that was probably called Einstein, *A Stone* or *The first Stone*[6].

---

[6] Pun: Einstein can be translated as a composition of the two words Ein, first and Stein, stone

Photoelectric effect

Einstein won his Nobel Prize in physics in 1905 and, contrary to popular belief, not for his *Theory of Relativity*, but because of *Theory of the Photoelectric Effect*, which he had reached from studies of electromagnetic radiation. This discovery gave him the right to become one of *"Quantum Physics"* founding father.

Consistent with his ability to open up to new ideas, he immediately accepted the mathematical simplification resulting from the use of the numbers made by Planck and used that theory to explain the radiation of the black body.

This new interpretation prevented, once again, the contemporary scientific community, always sceptical of the novelties, from accepting the study, which explained electromagnetic radiation as composed of energy packets or *quanta*.

Robert Millikan in 1910, five years later, confirmed definitively Einstein's hypothesis and also that the photon energy was connected to the frequency of radiation, with the experiment of the drop of oil that permitted to obtain the measurement of the electric charge of the electron.

Both Einstein and Planck had the ability to propose something so new that it goes counter-current to the knowledge of time. And this is pure genius.

Einstein's life is about a stimulating family and guides like Dad or Uncle Jakob or the tutor who has guided him safely. Innate stubbornness, rivalry with the school authority and a natural predisposition to science has completed the work. Some sources indicate that he undertook the study of differential and integral calculus around the fifteen years, along with the other foundations of higher mathematics.

I think it can be said, in general, that on an already fertile soil, properly cared for and with study programmes aimed at predetermined objectives, exceptional results would be achieved. At fifteen years old, our students normally start secondary schools and we have to wait for the most advanced university courses to come to balance the knowledge, especially mathematics that Einstein mastered at the same age.

Perhaps our school system is repetitive and the excessive redundancy of already expressed concepts is not calibrated to make optimal use of the potential talents of students.

The Atomic Models

We must be clear that when it comes to atoms, language can be used only as in poetry. The poet, too, is not nearly as concerned with describing facts as with creating images and establishing mental connections.

Niels Bohr About describing atomic models in the language of classical physics

The Handover

In those early years of 20th century, physical determinism, which had the diamond tip in the two theories of relativity, found a hard competitor in the probability of *Ondulatory Theory*.

The increasing number of experimental and theoretical tests on electromagnetism prompted us to interpret light radiation with a model based on electromagnetic waves similar to radio waves. However, there were still unexplained aspects, such as the nature of photons, which, due to the new optical experiments, left much to be desired. Some evidence continued to confirm that there could still be room to consider a corporeal nature and not just a wave.

The questions posed by the radiation of a black body appeared to have received almost all the answers, but Thomson's atomic model and Rutherford's subsequent model barely fit the wave theory alone.

It seemed in general that the planetary models of the atom were incompatible with electromagnetism, and therefore, there was a need for an additional effort of physicists to find a unification theory.

Unlike the physical method used by Einstein, it was the mathematical method that prevailed.

No longer experimental data to be interpreted globally in a theory, but mathematical formulas by which to justify a pre-packaged model that satisfied the experiments already carried out and all others that could have been conceived.

Douse this entailed the risk of designing only experimental tests in line with the theoretical model and not with the wide variety of events that occur in nature?

Or did it lead to another major inherent risk that is to use this method normally, thus preventing the growth of alternative theories that could lead to a more complete and more exhaustive theory?

And were these choices made with the certainty of their hard reality?

Would the achievement of the goals of the next physics have been conditioned by these experiments, which will be chosen and adapted only to the study of that single theoretical possibility?

Do we run the risk of paying a high cost because we will lose the ability to see beyond the limit of reason?

Does physics run the risk of a standardization process dictated mainly by mathematical laws?

Taking a leap in time and anticipating the answers with the knowledge of today, seeing how physicists are always persisting in the same search as the smallest or something that fits the Standard Model, which we will see in more detail later, it would seem that these doubts are largely justified.

It is since the 1970s that we have been unable to propose an idea that makes substantial progress.

But let us not anticipate the times and proceed gradually.

In addition to the physical discoveries, on December 24th 1906, after Marconi and Braun had made the wireless telegraph possible, which however consisted of impulse transmissions, Reinald Dessende made the first, true, radio transmission by devising a method that allowed, always through the electromagnetic waves used for the telegraph, to modulate an electrical signal, so as to transfer also the variations that were then reinterpreted upon arrival.

1911 Rutherford Father of Nuclear Physics

The academic world, which until then had been little inclined to change, was beginning to show great signs of vitality in the face of the new proposals that now arrived more and more frequently. Scientists were enthusiastic about the possibilities that were revealed thanks to the theories on the double nature of electromagnetic waves and the birth of *"Quantum Mechanics"*.

In 1911, Ernest Rutherford (1871-1937), British physicist from New Zealand, following his experiments proposed an alternative that resolved the contradictions of *Thomson's Pudding Model*, moving to a *Planetarium Model*, similar to our solar system: he imagined the atom composed of a central nucleus of positive charges with negatively charged electrons around rotating at a distance.

Occasional results were obtained: everyone contributed with what they found, without an organic development program, like many small pearls in a necklace that became more and more precious with each passing day with the addition of new elements. Rutherford was the paradigm of this advance, inexorable, but in small steps.

We recall that Becquerel in 1896, studying phosphorescence, had discovered that some uranium compounds impressed a photographic plate, even if it was not exposed to light. Later the Curies had found other elements that behaved in the same way, including radium, which gave the phenomenon the name of radioactivity.

In 1899, Rutherford, young researcher under the tutoring of J.J. Thomson, while studying the natural radioactivity emitted by various samples of materials, including uranium, had realized that the it emitted two types of radiation, which he called α *alpha* and β *beta*, and discovered that both were made up of charged particles of different mass of which he did not understand the origin.

These results allowed him to understand that the radioactivity was caused by subatomic transformations, even before having cleared his model of the atom that predicted the nucleus.

He was a very concrete man who made of simplicity a lifestyle, so much so that he loved to tell his collaborators:

> *I always believe in simplicity, being a simple man myself.*

For this reason, in order to understand nature and reveal its most intimate secrets, Rutherford tried to design simple and convincing experiments: he covered a sample of uranium gradually with more aluminium foils and measured the amount of radiation that pierced. With two sheets the radiation pierced, even if attenuated; by adding a third sheet there was no significant reduction and did not decrease even by adding the fourth and fifth layers. Only by covering the sample with many layers of aluminium the radiation disappeared completely.

What conclusion could be drawn from these observations? That uranium emitted two different types of radiation, one more penetrating than the other.

Thus Rutherford defined the weakest *alpha rays* and the other more penetrating, *beta rays*, from the first letters of the Greek alphabet.

Almost simultaneously, in 1900, working on *radium* samples, the French chemist and physicist Paul Villard had checked with experiments that there is also another electromagnetic radiation, more penetrating than the A and the B. And in the 1903 Rutherford, studying them in depth discovered that they were very fast and did not suffer the influence of magnetic fields and sensed the nature of electromagnetic radiation. He called them γ *rays*. It was later discovered that these rays can penetrate well into materials and that they have the ability to destroy chemical bonds.

Gamma rays are the main danger when working with radioactive materials and, unfortunately, it took many years and many victims before scientists realized these risks...

Subsequent reflective experiments on the surface of a crystal have confirmed the nature of *electromagnetic radiation.*

Ionizing Radiation	*Discovery Year*	*Description:*
		Ionizing radiation includes x-rays, gamma, alpha, beta and neutrons. They are particles or electromagnetic waves with high energy content capable of breaking the atomic bonds of atoms, releasing electrons and transmitting electric charge to neutral atoms and molecules. The atoms thus modified are called ions. Some of this radiation can

		cause biological damage.
Alpha Radiation	*Rutherford in 1899*	*It is an ionizing radiation of a corpuscular nature (particles) directly ionizing and with low penetration capacity. They are named after the first letter of the Greek alphabet because they were the first to be discovered. They consist of 2 protons and 2 neutrons bound by the strong force, hence He nuclei. They penetrate matter with difficulty and can even be stopped by a sheet of paper.*
Beta Radiation	*Rutherford in 1899*	*It is an ionizing radiation of a corpuscular nature (particles) directly ionizing, consisting of electrons or positrons emitted by the nucleus in a process defined as beta decay. They are high energy more penetrating than alpha rays and arise when a neutron decays into a proton. It is emitted by some types of atomic nuclei in a process defined as beta decay.* *This radiation is moderately penetrating and at least one*

		aluminium plate is required to stop them.
Gamma Radiation	Becquerel in 1896	*Gamma rays are photons emitted by the nucleus when an excited atom is de-energized by emitting a gamma ray. They are ionizing electromagnetic radiation. They propagate as waves or quanta of energy. They have a very high frequency and are very dangerous for humans. They irreparably damage the molecules of cells that begin to develop genetic mutations or can directly cause death. This radiation has no charge and easily penetrates matter. Requires lead bricks for shielding.*
X Radiation	Roentgen in 1895	*It is an ionizing electromagnetic radiation caused by electrons that change orbitals and propagate in the form of waves or energy quanta. They are very risky and easily penetrate materials and for this reason they require special attention and insulation by means of lead*

		plates or concrete walls.
Neutrons: among the ionizing radiations there are also neutrons. They ionize indirectly, either through the nuclei of the material they pass through or through the charged particles and photons they produce through nuclear reactions. They can make the affected atoms radioactive: this feature distinguishes neutrons from the ionizing radiation seen above.		

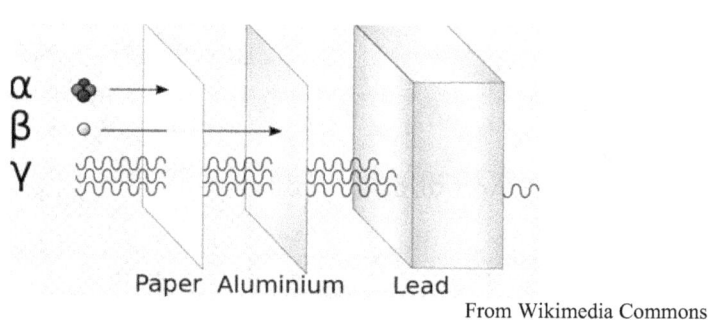

From Wikimedia Commons

Rutherford used the *alpha rays* so often that his colleagues said that he seemed to have created them specifically so that he could do his experiments best. The efficient techniques he had developed and, in some cases the luck, enabled him to understand the innermost secrets of atoms.

At the beginning, alpha rays, with their positive charge, were an enigma, while beta rays, with negative charge, were immediately classified as electrons. The other evidence that immediately appeared clear was that both of them did not behave like the known X-rays. It seemed as if they had a very small material consistency but with electrical charge and was emitted by certain materials called radioactive for reasons unknown at that time.

Rutherford developed the *scattering technique*, which became in the following years a standard experiment in nuclear physics and which allowed him to give many answers to questions related to alpha particles. He had no models or other studies to compare with other than the 1904 *Thomson Pudding Model*, and lonely had come to the conclusion that alpha particles were, in some ways, similar to helium atoms. He was sure they belonged to the same corpuscle because, being neutral atoms, if there was a negative charge evidenced by the presence of electrons, there had to be also a corresponding positive part that had to be something else always belonging to the same corpuscle.

Clearly someone else was beginning to ask the same questions and in the 1904, in an article published in Nature magazine, Japanese Hantaro Nagaoka (1865-1950) proposed a planetary atomic model, simulated on the model of Saturn and its rings which, although immediately bankrupt, had the goal of proposing a possible alternative: for Nagaoka the electrons rotated on concentric rings around a positive central charge. The model was unstable because it was explained using gravitational forces, therefore attractive and not repulsive, such as those Coulombian.

Rutherford, however, continued with various types of experiments using *alpha particles* and *scattering* and, from the results obtained, verified that the predictions were conditionable by multiple parameters, first of all electric charge and mass. Taking appropriate advantage in the preparatory phase of the experiment with the elements at his disposal, type of atom, corresponding, known mass, known charge, etc., he realized that he could predict the possible diffusion would be and, only afterwards, with a reverse process on unknown starting elements, apply the same mathematical laws to determine the initial parts of what he did not know.

It was now clear that the scattering, or rather the deviation, or rather the change, as it is best meant to understand it, compared to the initial conditions was produced only by a single collision that could have occurred only if much of the mass of the atom was concentrated in its center, and this was a good starting point to reveal something else that the atomic structure of matter still concealed.

Johannes Wilhelm Geiger (1882-1945), a trusted collaborator who had collaborated in the experiments for many years, later told that in December, 1910 Rutherford had suddenly entered his room and told him that he had understood how the atom was made by the way the alpha particle scattering presented itself.

It is evident, in this case too, that the intuition of the scientist, of a visionary man, is different from the intuition that a technician or a pragmatist of the rules can have.

Rutherford's incredible proposal is to be contextualized in the period he was in, when few spoke of atoms and those few knew absolutely nothing about how truly representative he was. Today an atom is quietly drawn on school notebooks even by children. So Rutherford, going against normal common sense, imagined the atom as a nucleus, relatively large in mass compared to the electron and electrically charged, surrounded by a space for much of the vacuum, similar to the solar system.

To understand the dimensional scale we are talking about, let's imagine a stadium with the nucleus in midfield with the size of an apple seed: the electrons are smaller dots distributed on the stairways of the spectators, the entire mass is almost completely concentrated in the apple seed. .

However, although the planetary model for Rutherford was now a clear vision, he still could not understand what happened to the electrical charges concentrated in the nucleus.

March, 1911: in a letter to a colleague, he hypothesized that the central nucleus could have a positive charge, thinking that the alpha particles, also positively charged, behaved in the vicinity of the nucleus as a comet would do on its way around the Sun.

May, 1911: Rutherford published a "beautiful and famous article" about scattering, as it will be judged in retrospect by Heilbron and, among other things, he wrote:

> *Considering the evidence as a whole, it seems that the simplest thing is to assume that the atom contains a central charge distributed in a very small volume.*

Keeping in mind the null knowledge of that period, Edward Neville da Costa Andrade (London, December 27, 1887 - June 6, 1971), English physicist, judged him among the most significant scientific articles ever, defining the contribution given by Rutherford as:

> *the greatest change in our ideas of matter since the time of Democritus [...] four hundred years before Christ.*

Rutherford became more and more convinced that the nucleus, in which almost all the mass was concentrated, of positive charge, was in turn composed of several sub-elements[7],

---

[7] They will later be called *protons*

some of which were subsequently called protons because at that time they did not yet have a name. He was also able to calculate that these would have a mass of 1836 times that of the electron, based on the expulsion capacity of the alpha particles.

He also established that in an atom the nucleus was about 100.000 times smaller and that, its internal volume, was almost completely empty, because the mass was almost compacted entirely into the nucleus. This empty space could not be filled by electrons from other atoms, perhaps because of electric fields so powerful that they rejected electrons from other atoms.

Rutherford, always studying the experimental data of scattering with various elements, also understood the existence, in the nucleus of the atom, of something else, probably another particle that contributed to varying its mass, which initially defined a neutral proton which mass was equal to the proton but with neutral charge. The instruments and experiments available at that time were able to find protons and electrons, but not the other particle. Only in the 1935's James Chadwick (Bollington, 20th October 1891-Cambridge, 24th July 1974), an English student of Rutherford, proved the existence of these particles.

By now the atomic models on the scene of physics had erased the hypothesis of indivisibility postulated by Democritus: the elementary bricks of matter also had a description of the inner parts and their structure. The atom turned out to be divisible and the nucleus, in turn, was composed of subparticles.

Rutherford's model was an important turning point: it provided answers to so many doubts that could not be solved, and among others it made it clear that alpha particles were positive-charged nuclei that, once expelled, could slow down and attract electrons by becoming electrically neutral as common helium atoms.

In this case too, as will often happen later, Rutherford's hypothesis did not spread widely among colleagues of the time and generally in the scientific world and himself did not promote it with the necessary conviction, making only two brief references to it in an article published almost two years after the discovery.

We can now say that Rutherford's model was of little importance without proper recognition.

In the 1937, year of his death, the New York Times obituary honored him with these words:

> *Few men are allowed to achieve immortality and even fewer are those who are allowed to reach Olympus while they are alive.*
>
> *Lord Rutherford succeeded in both.*

1913 Bohr: Einstein's Opponent and Friend

Although Rutherford's model was conceptually innovative, it did not include much of the information that was already available on the physics of the atom. In fact, as had already happened for Nagaoka Hantaro (Umura, 1865-Tokyo, 1950), his fact-testing model could not work perfectly because, this time, mechanically unstable: the rotating electron should have dissipated electromagnetic energy until it fell on the core, causing the instability and destruction of the atom itself and consequently making critical the mass of matter it constituted.

In the 1912, the Danish physicist Niels Henrik David Bohr (Copenhagen, October 1885-Copenhagen, November 1962), Manchester while living in Manchester, studying the light spectrum emitted by atoms and verifying that it was different for each type of atom, suspected that it was an effect of the atomic structure of the element under examination. Then decided to confine himself to studying the spectrum (Balmer series) of the hydrogen atom, which was quite simple and to reinterpret its understanding, introducing *Planck's constant h* and Einstein's theory of energy quantification into mathematical formalism, applying them later the results on Rutherford's atomic model.

To do this, he formulated three hypotheses that took his name:

Bohr's first hypothesis	*electrons revolve around the nucleus in circular orbits without emitting electromagnetic radiation. Each of these orbits has a specific radius.*
Second hypothesis	*electrons have an angular momentum, quantized, which can only have integer*

	values and multiples of $\hbar$, where $\hbar$ is the Planck substance divided by 2π. It cannot have intermediate values between those allowed.
Third hypothesis	an electron does not change its energy state as long as it remains on its own orbit and on the first orbit (ground state) it has the least amount of energy. To jump between the orbits towards the outside (move away from the nucleus) it must receive a fixed amount of energy that corresponds to the energy difference between the two orbits. Conversely, if it wants to move from an outermost to an innermost orbit, it emits an amount of electromagnetic energy equal to the energy difference between the two orbits. This explains why the bands of the electromagnetic spectrum are discrete and not continuous.

In 1913 Bohr was able to propose the first quantum model for the hydrogen atom where electrons were, only, at well defined and not random distances, multiple of the radius.

Bohr's model, so conceived and completed by Planck's constant, now known by the name of *quantum*, immediately proved applicable to many elements and not only worked in this case, but also justified a whole series of other observations including the spectrum of hydrogen. All operating problems inherent in the stability of the atom of Rutherford's still unresolved theory were momentarily set aside with a typical Bohr-style solution: the atom emits or absorbs radiation only when one electron is excited and jumps from orbit to orbit between stationary states, otherwise there's no power loss.

Harry Moseley explained later that X-rays were emitted from the electrons of the inner orbits of the elements and that Rutherford-Bohr's planetary model adapted satisfactorily to experimental testing.

Having satisfied all the assumptions of the model, which imagined the rotation of electrons on circular orbits well defined by the laws of energy quantification and, given that atoms in quantum leaps absorb or emit energy, Bohr concluded that the treatment of mechanical and electromagnetic quantum phenomena could not be independent of a double corpuscular and wave nature. All this was established in the principle of complementarity, presented to the International Congress of Physics of Como in 1927, so that the dual aspect of some physical representations of the phenomena at the atomic and subatomic levels, one corpuscular and the other ondulatory, both true and mutually complementary, cannot be observed simultaneously during the same experiment.

This principle was then taken up by Heisenberg and became fundamental to the *Uncertainty Principle*.

This path of meaning seemed to give the hoped results and the improvement of the instrumentation and spectroscopic techniques led to additional clues that served to increase the experimental results more and more. In the analysis of the spectrum the resolution increased and it was seen that each spectrum line was in turn composed of other series of very close and thinner rows, thus assuming that there were sub-levels.

The model that Bohr perfected had certain parameters and, therefore, rules were introduced for the coding of the value n

which identifies the permitted energy levels and, which had to be understood, consistent with the experimental results of the spectrum, between 1 and up to 7 and assume integer values.

The sub-levels, highlighted by the thin lines, were quantified and for their description it was necessary to consider the introduction of additional quantitative numbers.

The Bohr atomic model proposed in 1913 worked correctly for the hydrogen atom that has only one electron and a very simple spectrum to study. It still functioned partially for some alkaline metals such as lithium and sodium, but did not provide answers for poly-electronic atoms that had a more complex spectroscopy, such as helium that had unscheduled lines.

The inadequacy of Bohr's model was due to the use of the laws of classical mechanics he had used, which were not suitable to explain the properties of subatomic particles such as electrons, protons and atoms.

Bohr Heirs

Arnold Johannes Wilhelm Sommerfeld (Königsberg, December 5, 1868 - Munich, April 26, 1951), German physicist, in 1915 made some changes to improve the model by changing the orbits from circular to elliptical and moving the core to occupy one of the two focuses. Thus exploiting the rules on the geometric places of the ellipse, the electrons passed to have 4 quantum numbers, 3 of which were used to identify the orbit and 1 the mode of rotation of the electron. The next step was to imagine that on the same orbit there were two electrons at the same time but with opposite direction of rotation, with magnetic fields of opposite sign but of equal intensity, so as to avoid problems due to repulsive forces. In this way we began to understand the concept of opposite spin for electrons.

This whole construct stood up, but they were artifices that solved many problems without a reason documented by a solid theory. Nature was ignored: how could all these hypotheses be in agreement? Was there a still unknown law that contemplated the result of observations for each known atom that was taken into consideration?

It was by no means possible to think that some atoms behaved one way and others behave differently.

This period of uncertainty lasted until Pauli enunciated his principle, according to which two electrons with the 4 equal quantum numbers cannot exist in an atom.

1919 The Proton

In 1915 Einstein presented his *Theory of General Relativity* while atomic physics seemed to have stalled, because classical physics was not sufficient to explain all the experimental tests and the available mathematics could not provide consistent results. The time was not yet right: it would take another ten years to develop the tools capable of better explaining *"Quantum Theory"*.

An obligatory pause in scientific research and beyond was caused by the First World War which tore Europe apart from 1915 to 1918. Many scientists had to abandon research to devote themselves to more pragmatic objectives, not least that of going to fight for their country. In any case, atomic research was no longer able to progress. Finally on June 28, 1919, the Treaty of Versailles put an end to the Great War but among the many victims there were also many young scientists.[8]

In 1897 Thomson discovered the electron. In 1911 Rutherford discovered the core. In 1932 Chadwick discovered the neutron and physicists were definitively convinced that in order for the atom to be considered electrically neutral, as indicated by all the experimental results, there had to be constituents with a positive charge, which had to be in the nucleus.

There was no discovery directly attributable to a single mind. Instead, it was a gradual process to which many contributed, each making his own little discovery, but Rutherford is rightfully the

---

[8] Among the many curiosities that took place in those years, I like to point out that, for the first time, a state intervened to tax a fuel. This primacy belongs to Oregon, which imposes a tax of one cent per gallon for the good of the American nation. Since then, the politicians will never stop.

one who unequivocally fixed the description of the proton in the history of physics.

> *Rutherford was firmly convinced that there were subatomic particles and this led him to discover the "alpha and beta rays" and to understand radioactivity. The discovery of the "proton" can also be attributed to Rutherford himself in 1919.*

In 1911 Rutherford's discovery of the nucleus had shown that these positive charges were concentrated in a very small fraction of the volume of atoms.

In 1919 the scientist discovered the possibility of changing one element into another by bombarding it with *energetic alpha particles*, which today we know are helium nuclei: this phenomenon is *transmutation*, which has always been coveted by the alchemists of past centuries.

Thus, in the early 1920s, physicists carried out many experiments with the newborn technique of *transmutation* by practicing on many atoms and discovered that, again in this process, hydrogen nuclei are emitted: it could be concluded that the hydrogen nucleus had a fundamental role in the atomic structure and that the positive charge of the nuclei could be explained by an integer number of hydrogen nuclei.

It was Rutherford himself in 1920, referring to hydrogen nuclei, coined the term *proton*, from the ancient Greek *proton*, which means *first*, thinking that they were the elementary constituents of the nucleus of atoms. From these experiments he also deduced the presence of another particle which, in 1932, under his supervision, was discovered by his pupil James Chadwick, who also won in 1935 Nobel Prize for it. The new particle was the *neutron.*

Only in the early 1960s, thanks to the most sophisticated experiments of the Stanford Linear Acceleration Center and the National Scientific Research Center on nuclear physics in Germany, it was possible to ascertain that the proton has a complex internal structure and, therefore, it is not an elementary particle.

Even today, according to CERN researchers, *much remains to be discovered about this particle.*

Particles Exist ...

> *I have done a terrible thing. I have postulated a particle that cannot be detected*
>
> *Wolfgang Ernst Pauli*

1923: The Compton Effect

Arthur Holly Compton (Wooster, 10 September 1892 - Berkeley, 15 March 1962), an American physicist, in 1922 among the many experimental confirmations, through the discovery of the Compton Effect, verified unequivocally, that the electromagnetic radiation in some circumstances was composed of corpuscles, thus definitively confirming Einstein's 1905 theory of the photoelectric effect. It took fifteen years for Einstein to be right.

Let's see how this experiment works: imagine we are in an empty room and scream at a wall which, reflecting the sound waves generate an echo. Compton decided to replicate the mechanics of this experiment by replacing sound waves with X-rays and found that these, reflected by the electrons, were energetically weaker and of lower frequency. It is as if the echo had responded with a lower pitch.

This behavior did not fit well with known wave patterns of electromagnetic radiation, since it seemed that part of the momentum of an X particle passed to the electron like two billiard balls, so that the stationary one hit by the moving one acquires part of the energy by bouncing according to the angle impressed by the incoming ball.

For the classic wave model of X-rays and electrons there should have been a transmission of the oscillation at the same frequency as the incident wave with the simultaneous emission of a radiation always at the same frequency.

In this way Compton explained the observed events by considering the nature of X-rays as a particle, using Einstein's hypotheses and confirming their existence.

For Compton, as already said, these particles were like billiard balls with a momentum that in the elastic collisions against the electrons transmit part of their energy, even being partially rejected.

Diffraction and interference confirmed the wave nature and the experimental confirmation carried out by Compton also granted a corpuscular nature to the light which thus had both characteristics.

In that historical moment, with newly acquired and so opposing knowledge, it was a real leap into the void: who would have been willing to venture on what seemed to be such a risky and contradictory path?

1924: Louis de Broglie and the Duality Wave

The 1920s were frantic for physics: discoveries followed more or less significantly. Many experimental confirmations of previously exposed theories were obtained; many of the new proposals made the general picture clearer in some cases, but in other cases made a fuss. The post-Rutherford time was therefore a pioneer of novelty but also very confusing.

It is necessary to put order to understand the correct sequence of events but also to understand the real nature of the atom that begins to assume a definitive configuration, without being influenced by either graphical representations or theoretical interpretations, which were always reductive of the true configuration and not always adhering to the true essence.

Besides, this is the limit of a *model*.

Louis de Broglie was a young French aristocrat, with well-groomed moustaches and great elegance in his clothes. One of his ancestors was named Duke by Louis XV in the 1742 and Louis could also take the title.

Also De Broglie was taught privately by tutors who went home. He soon became interested in politics as well as family tradition but, when his father died, he was directed by his brother, who had become the head of the family, towards university studies with a humanistic orientation that did not, however, excite him excessively.

While attending his brother's science lab, he became passionate about X-ray physics. He tried, for the first time in vain, to take a physical exam. He was a very private man, who loved to devote himself to his interests and his brother Maurice will report that he did not love leaving home to continue his studies.

A fortunate coincidence made him definitively passionate about physics: his brother Maurice, that was a notable of France, was appointed secretary at the first Solvay Congress in October 1911 and was accompanied to Brussels by young Louis, who, however, remained at a distance from the eminent scientific personalities there.

When he returned home, Maurice told him about the events he had been listening to, telling him about the debates and what he had tried to be close to people of Einstein's level, Marie Curie, Planck and Rutherford. The young Louis was thus conquered by his passion for physics and in the 1913 he obtained the "Licence ès Science", a certificate of knowledge of physical disciplines.

Unfortunately, the Great War was at hand and the young man was recruited to serve in Paris...

In August 1919, deeply disappointed at having lost six long years to the war, Louis was discharged and immediately resumed his passion for physics.

His brother Maurice, however, had kept him updated on the results achieved by experimental physics and in the early 1920s Louis, becoming passionate about understanding the dual nature of light, published articles on X-rays.

Confronting each other, the two brothers became convinced that to explain the nature of light there was a need for an explanation that integrates both the wave motion and the corpuscular nature, since the two theories were confirmed by the diffraction and interference experiments and the photoelectric effect. .

In 1905 Einstein had won the Nobel Prize by explaining the photoelectric effect as the emission of electrons by a metal subjected to certain frequencies of light which was composed of corpuscles, photons.

De Broglie, even before the publication of Compton's results, was already sure that there was something similar to atoms in the light rays and already in his mind the possibility of a dualism coexisted.

Not only that, but he was also struck by another idea which he described as follows:

> *After long reflections and meditations in solitude, suddenly, in 1923, the idea came to me that the discovery made by Einstein in 1905 could be generalized and extended to all material particles, in particular to electrons.*

In 1923, starting from the results that confirmed that only electromagnetic radiation, light, had both ondulatory and corpuscular behavior, reversing the point of view he proposed in his doctoral thesis the theory that electrons also behaved like waves; then he extended this dualism to any material particle, coming to the conclusion that double nature was a universal property of matter.

De Broglie's hypothesis stated that each moving particle is associated with the typical physical properties of waves and that by assigning a certain frequency of starting wavelength, this became a *pilot wave* able to adapt exactly to the predicted orbits, in such a way as to satisfy the arrangement devised in Bohr's quantum atomic model and, he defined the wave it emitted as a *matter wave*.

In this way, the unsolved questions of Bohr's model were also answered and his equations gave a fundamental impetus to the development of "Quantum Mechanics".

de Broglie received the Nobel Prize for his contribution in 1929 and thus spoke of himself:

> *... having much more the state of mind of a pure theoretician than that of an experimenter or engineer, loving especially the general and philosophical view ....*

Classical physics, with its reductive laws, had not until then allowed to predict and follow the behavior of each electron, but after de Broglie it seems that a turning point was looming on the horizon.

It was coming the time that would change the physics of the atomic world definitively: it was Schrödinger who proposed *"Quantum Mechanics"*, introducing laws of a probabilistic nature that provided a mathematical and conceptual setting more suitable to explain the wave behavior of electrons...

1925 Pauli Exclusion Principle

Wolfang Pauli, who was not exactly one of the funniest scientists, continuing to have to do with his beloved *"Quantum Physics"* and other colleagues, at some point in his life seems to have declared:

Physics is now too difficult. I'd rather be a comedian or something rather than a physicist.

While de Broglie had managed to fix Bohr's logic of the atom by formalizing the double nature of subatomic particles, in 1925 the Austrian physicist Wolfgang Ernst Pauli (Vienna, 25 April 1900 - Zurich, 15 December 1958), Austrian physicist, contributed to the strengthening of the system of *"Quantum Physics"* with its *Pauli Exclusion Principle* which allowed to rationalize the positioning of electrons on the orbitals. Two electrons, or even other *fermions*[9], cannot share the same quantum state in the same atom or molecule. That is, no pair of electrons in an atom can have the same *quantum numbers*[10] which were first established by Bohr. Thinking thus, on each orbit we can find at most two electrons, but with opposite spin; each electron is characterized by four different quantum numbers. Affirming the uniqueness of each electron is also equivalent to stating that,

---

[9] In quantum physics, fermions are the elementary constituents of matter such as electrons, which structure and make atoms impenetrable, hindering their neighbors, and bosons which are elementary particles that allow the action of three of the four fundamental forces of nature.

[10] Each electron in an atom can be uniquely identified by a set of four quantum numbers.

Advanced mathematical skills would be needed to explain their function, but they are all conventional numbers: three of them can only assume simple integer values; the fourth is a simple fraction.

according to the *Pauli Exclusion Principle*, two or more electrons that have all four equal quantum numbers cannot coexist in an atom. What does this we have just said conceptually mean? And what are quantum numbers?

Let's take an example and try to understand what kind of atom representation we have finally arrived at, because with Pauli we complete the picture that describes how all atoms are made.

We can imagine an atom like a building and electrons are its condominiums.

To date, 2021, the periodic table is made up of 118 elements.

Each element is a different building type, so there are 118 different building types.

The most common elements in nature, such as iron, oxygen, magnesium and so on, are the types of construction of the various buildings: same element, identical buildings; two different elements are two types of different buildings and so on.

Quantum numbers are like the construction and housing rules that the masons who build the building must adhere to and later like those that the condominiums will have to respect when they go to live there.

Each element has the same quantum numbers, so each type of building respects the same construction rules. The quantum numbers that identify the electron in each atom are four; it follows that the building will have four types of characteristics that we will define as follows:

Characteristic of the Buildings	Quantum Number	
The number of floors can be from 1 to 7 maximum	n	Quantum number main *n*, defines the number of allowed energy levels, max 7
Each floor of the building can have 4 different types of apartments: economic, standard, elegant, furnished.	ℓ	defines the type of orbits allowed for each level, s p d f
Based on the types of apartment we establish, how many there are per floor: so we will have 2 economic, or 2 economic and 1 standard, or or 2 economic, 2 standard 1 elegant, etc.	m	magnetic quantum number m, defines the number of orbits allowed for each type
This last parameter is the most exclusive: the inhabitants of each single apartment are of the opposite sex or male or female or a man and a woman.	m_s	spin quantum number identifies the two electrons of the orbit, if any. By the Pauli principle, in each orbit there can be a maximum of two electrons with the controversial associated spin vectors or, according to nomenclature, antiparallels.

It becomes much easier to understand how the elements found in nature, listed in the periodic table, have properties that can be described with the apartments of the individual buildings in the example described above. The properties of the elements depend on the configuration of the electrons of the atom that have been formed according to a precise model.

To refer to the example of buildings, it is as if the tenants to go to live there must respect the rules for filling the apartments.

To proceed with the ideal "construction" of the atoms it was decided to follow both the *Aufbau Principle*, a system consisting of three steps that fills from the bottom up, and with the *Pauli Exclusion Principle*, so that in an atom there is no there will never be two electrons with the same quantum numbers.

Finally, due to *Hund's rule*, or *Principle of Maximum Multiplicity*, if more orbitals have the same energy, then the electrons are arranged first of all one per orbit and, only if their number allows it, they subsequently saturate the other orbitals.

1925 Heisenberg and the Matrices

Heisenberg, a pillar of *"Quantum Physics"*, associated himself with Pauli's way of thinking and commented that

> *Not only is the Universe stranger than we think, it is even stranger than we can think*

Werner Karl Heisenberg (1901-1976), was a German theoretical physicist Nobel Prize for Physics in 1932. From 1920 to 1923 was a student of Sommerfeld, of Franck, Born in Vienna and of Hilbert at the University of Monaco and Gottingen. It was Sommerfeld who understood its capabilities and directed it towards the study of atomic physics with the clear intention that its capabilities would help Bohr and the newborn *"Quantum Physics"*.

The increasingly obvious contradictions between the theory of the quantum developed by Bohr and then by Sommerfeld, the classical electromagnetic theory of light and the enthusiasm of the new groups of physicists who attended his frequent trips to Munich, Gottingen and Copenhagen, was among the reasons that helped to convince the young Heisenberg and his colleagues that the theory of *quantum* could have been an alternative answer if a new *"Quantum Mechanics"* had also been devised.

Until then the concept of atom and its sub-elements, nucleus and orbits of particles had been associated with a system similar to solar orbits and their motion of planets. The new idea, however, came from the conviction there was a need to think of other mathematical instruments that could explain the results and properties revealed by the atoms quantified in the atomic experiments that were multiplying in universities and research centres.

This intuition was the starting point for the development of the new *"Quantum Mechanics"*. Copenhagen with Bohr and Pauli, and Gottingen with Born and Pascal, pressed for all those who were able to intensify their studies in a joint effort. It was in July 1925 that Heisenberg came to a first conclusion which was not very convinced himself.

It was Max Born who understood the potentialities of this new proposal: that of applying an abstract branch of mathematics, *the theory of matrices*, to the interpretation of the results they possessed. So in 1926, along with *Pauli's principle of exclusion* for the construction of atoms, Heisenberg and his group presented their *"Quantum Mechanics"*, based on the calculation of matrices, which was able to predict and explain many atom behaviors.

The scientific community, as usual sceptical of a radical novelty, had many difficulties in accepting the new proposal because it was highly abstract and mathematical, despite the undeniable results achieved by this method.

The formalism on which the *Theory of Matrixes* was based was certainly complex, and so abstract that even today, knowing through other theories what happens in the atom, it is difficult to describe even the simplest properties using this system, as well as understanding what happens in the atom.

In any case, the pillars of classical physics are now almost destroyed, the idea of trajectory no longer makes sense and the position and velocity of a particle can no longer be determined simultaneously. The new mathematics and physics move from a certain and visible determination towards the indefinition of probability, also feeding a philosophical interpretation of the apparent illogical discoveries.

Dirac opposed this way of seeing things, a staunch defender of the goodness of relations with classical mechanics, who continued to advocate only the change of the mathematical characteristics of physical quantities that satisfy these relations.

1926 The Compton Photons

In 1905 Albert Einstein explained the emission of electrons in *the photoelectric effect* using the ideas that Max Planck had in 1900 to explain *the black body radiation* and he used for the first time the concept of discrete energy packets, the *quanta of energy*, in a completely different theory from what they were meant for.

In 1923 Arthur Compton used the same method, using *energy quanta* to explain why *X-rays* had corpuscular nature.

On the other hand, in 1926 Gilbert Newton Lewis (Weymouth, 23 October 1875 - Berkeley, 23 March 1946), an American chemist, was the first to definitively clear these *energy quanta* even in the case of light rays and used the term *photon* (from the Greek *phōs, phōtos*, "light"), to define the single unit of *Einstein's quantum of light*.

The *photon* has a life which is believed to be infinite at the moment because, not being a mass carrier, it never decays.

The interaction between particles can generate or destroy the photon, but this cannot decay without external intervention; it has an energetic charge without possessing mass and is affected by gravitational fields.

It should be specified that in the *Standard Model*, which we will see later, it is also possible to calculate the possible mass of the electron but there is no experimental evidence to confirm it. At least this is the situation today, year 2021.

Its peculiar characteristic, that we assumed to be true is that of being equipped in the vacuum with a speed equal to that of light, while, if it passes through matter, the interaction with other particles gives it mass and its speed decreases under the value of "c", that is the letter that identifies the speed of light.

The emission frequency of the radiation of the *quantum of light*, therefore of the *photon*, is its energy, which can belong to many categories of radiation, such as energetic X-rays, gamma, visible light, radio waves, infrared, etc. etc.

Photons belong to the *bosons* family and have no electric charge or mass and, since they are massless, they also have no spin units; they are thought to contribute to the electromagnetic field and therefore many believe they are field particles.

Basically physics does not yet have sufficient data on photons for a deeper understanding and therefore they are used as a mathematical formalism to justify the models that contemplate it.

All this shows us that we still have a long way to go before reaching a total understanding of the Universe that surrounds us.

... or They Probably Exist?

> *Since the mathematicians have invaded the theory of relativity, I do not understand it myself anymore*
>
> *Albert Einstein*

1925 Thomas Kuhn and The Events onwards

Speaking of Einstein and other exceptional minds, it still surprises today how it is possible to come to understand theories so different from the usual ones. It is clear that part of the explanation had to be in the mental processes that were activated in the scientist's head, in his personal history, in the circumstances that generally led a person like him to achieve results of such great importance.

I intentionally used the verb "bring", because I like to imagine that there is something, that some call intuition, other inspiration, or luck, that takes you gently into the palm of his hand and leads you to get these unusual and amazing results.

If we want to understand what happened after 1925 in the sequence of innovative proposals of atomic physics as if we are witness of those events, we also need to understand what was driving the entire race.

Thomas Kuhn (1922-1996) was an American physicist, historian and philosopher. In 1962 he wrote *Structure of Scientific Revolutions* in which explained how science changes over time, even if it does not always evolve, thanks to the "movements" he defined *paradigm changes*.

For Kuhn, a scientific revolution, well identified by a historical period or a new theory, occurs when there is an impossibility to arrive at an explanation of the new experimental results using those paradigms universally accepted until then; the paradigm is not limited to the theory until then used for habit, but incorporates the total perception of your reality with all the events that result.

Kuhn argues that the inadequacy of the old paradigms to explain the new results is circumvented by an extension of the acceptable meshes or, more commonly, by leaving out or ignoring the obstacle. Let us contrast these interpretations of Kuhn with what happened in physics in the passage after 1925: the warnings of a radical change taking place are felt, but some scientists wanted to maintain the status quo of a physics that provided perceptible results despite the finding of the inexplicability of new events that unequivocally showed a new physics on the horizon that, among other things, was necessary.

When the current paradigm is called into question by different results accepted as true, the rules in use until then go into crisis. This state of discomfort leads to the proposition and experimentation of new ideas, which could previously also have been put aside, until the constitution of a new paradigm and its new proselytites, with consequent intense intellectual comparison with the resistance opposing the followers of the old paradigm.

This is what happened after 1925: the vision of the quantum mechanical world, the new interpretation of Einstein's relativistic world, and the new picture that unfolds on the horizon of a probabilistic physics that will be definitively proposed by a trio of attacks such as Heisenberg, Schrödinger and Gödel, it could not have no consequences on the nervous systems of individual scientists who, let us not forget, are not men of "peace", rather, are men willing to defend their integrity and intellectual honesty with nails and teeth.

These disputes generated an infinite series of "attacks" and "counterattacks", either with empirical data, with rhetorical or philosophical arguments, or even with mental experiments, until an apparent consolidation of probabilistic theory on other proposals.

In fact, the evaluation and importance of the proposals of the new interpretation of *"Quantum Physics"* is adapted to the results obtained according to what you want to see: scientists, in the end, are not only men but also gamblers who always and only want to win.

The simplicity of the possible mathematical formalism of Einstein's equations and other deterministic scientists, for many, did not have enough attraction to adapt to possible new theories.

Normally it is believed that it is time and human contribution that bring order: in our case from 1925 a boa ride has been made that still continues with the *Standard Model* to govern the subatomic world, with threatening reflections also on a possible cosmological understanding.

Max Planck argued that:

> *A new scientific truth does not triumph by convincing its opponents and making them see the light, but rather because its opponents eventually die and a new generation grows up that is familiar with it. . . . An important scientific innovation rarely makes its way by gradually winning over and converting its opponents: it rarely happens that Saul becomes Paul. What does happen is that its opponents gradually die out, and that the growing generation is familiarized with the ideas from the beginning: another instance of the fact that the future lies with the youth.*

Thus the change from one paradigm to another, what winners believe to be a scientific revolution or a paradigm shift, is nothing more than the sad conclusion of a long selection process, where not the best theory ends up being right but the one whose proponents survived. And we don't know if they survived because the theory is better, but only because the opponents had to give up because, in the end, they materially became extinct.

This principle is valid in every field and in every dispute and we could summarize it by noting that, in the end, history is always written by the winners.

1926 Schrödinger and Born

The changes that took place from 1925 to 1930 were mainly due to the incredible contributions of Heisenberg, Schrödinger and Gödel that with their ideas proposed theories so radical to upset the way of thinking about nature and reality.

It is suffice to say that much of the technology we currently use has its roots in the ideas of those men of that period.

Let's go in order and try to understand what happened.

Bohr's conceptions, which was perfected by de Broglie, seemed to partially satisfy the ideas of nascent atomic physics because, unfortunately, they had no real empirical foundation but were only the result of mathematical abstractions. And they were still partial, because they were unable to incorporate all the variables involved. At least until a new protagonist appeared on the scene.

Erwin Rudolf Josef Alexander Schrödinger (Vienna, 12 August 1887 - Vienna, 4 January 1961), was an Austrian physicist who won the Nobel Prize for Physics in 1933. He wanted to help avoid the mathematical difficulties that arose from the Heisenberg Matrices and, in 1926, proposed a mathematical function known as the wave equation: this allowed valid results to be obtained without serious calculation difficulties or conjectural deficiencies and fully described the behavior of subatomic particles in a global context of the whole subatomic world known up to then..

The theoretical changes that eventually led to the comparison between the two schools of thought, the one before Schrödinger and the one after him, were preparatory to the intuition of

Heisenberg's Uncertainty Principle and to the formulation of the Copenhagen Interpretation.

The most significant change consisted in replacing the quantum mechanical orbits predicted by Bohr and Sommerfeld with wave functions that obey the quantization laws we have already seen for energy.

Schrödinger hypothesized that electrons were waves and so they do not move, transmitting motion only through oscillations that are similar to standing waves. Whenever you look for an electron it is possible that you find it in a different place without the need to look for a movement: it is not important the particle motion but the amount of associated energy.

So no more orbits, paths on which electrons run as the planets of the solar system do, but energy levels, at predetermined distances, around the nucleus.

The electron, in the transmission of energy at very high speed that it makes on the path traced by the circular orbit around the nucleus of the atom at whatever level it is, generates waves that propagate and that we detect with our observation instruments. These waves, such as, can be mathematically interpreted with a mathematical function.

The other idea that led Schrödinger to success was the introduction of a new mathematical operator, neither more nor less than a product or a root, which linked the wave function to its change over time: the *Hamiltonian Operator*.

This description of a quantum system, as predicted by Erwin Schrödinger's studies using the wave function, is still widely used today.

The study of the motion of the electron, alone had required the birth of a new branch of physics, *"Quantum Mechanics"*, and its probabilistic approach was the only one that could explain it even if with the great logical contradiction between particle and wave. It was one of those paradigm shifts that Kuhn referred to with all the sometimes contradictory implications that this implied: in the end which of the contenders had proposed the most correct thesis?

Physicists wondered how mass and electric charge, peculiar characteristics of particles, could coexist with the two characteristics of waves, wavelength and delocalization. Not only that, but they wondered if all the known figures in classical mechanics, which the waves built, were also satisfied by two assimilated particles with a wave nature and, in particular, if the phenomenon of interference was preserved.

Davisson Germer Kunsman in the United States and G.P. Thomson (son of J.J. Thomson) in Aberdeen in Scotland, in 1927 experimentally verified that, in fact, electrons emitted and behaved like waves and as waves produced diffraction and interference patterns: so Schrödinger's claims were confirmed.

This confirmation raised great debates and opened a new window to a metaphysical understanding of subatomic phenomena for the great implications that it implied. Scientists were now stunned and physicists were no longer surprised at anything. Schrödinger, to develop his equations, used the *Theory of Waves of Matter* intensively, uncovering an authentic Pandora's Box: matter was no longer such.

Schrödinger defined these wave functions with the term *orbitals*, which are nothing more than mathematical functions that expressed probability distributions, that is, the areas of space where the electron was most likely to be found. For this reason the new atomic model was called the orbital model.

If you want, take a look at a beautiful animation created by the great Italian designer Bruno Bozzetto for Piero Angela's Quark program. It is very clear and fully explains the model of *"Quantum Mechanics"*.

Max Born (Breslau, 11 December 1882 - Göttingen, 5 January 1970), a German naturalized British physicist who won the Nobel Prize in Physics in 1954, in 1926 proved that the result of the wave function could not be deterministic, but was itself a probability function: the result produced another equation that described the probability that an electron could be in a certain region of space around the nucleus of an atom with no indication of its actual position.

An irreversible leap had been made: the mathematics and mechanics of *"Quantum Physics"* became comprehensible only to a few truly specialized workers.

The new equations did not provide numerical solutions but generated other functions, called *wave functions*, which in turn generated other functions that described the possible states of the electron in its existence around the nucleus of the atom.

The determinism of classical physics, where everything was perfectly predictable, gave way to the probabilities of *"Quantum Physics"*, in which we know nothing of what is happening but we can only predict possible events...

Schrödinger's equations are considered by many to be postulates[11] of *"Quantum Mechanics"*: normally the postulates

---

[11] Postulate: a postulate is an affirmation of a principle, impossible to prove, and to accept because it produces valid results to explain certain facts or to produce a theory. To understand better the postulates are useful because they predict and satisfied experimental evidence.

are verifiable with paper, pen, tape measure and observation. In this use, however, we must believe and set all subsequent physics on something that we are not able to verify because it is probable.

The obtained results only partially confirm the theory, which is so always incomplete. The question is: are they sufficient to demonstrate their validity or are they self-referential, because they produce results that can only be certified by the same theory that they should instead explain?

The structure, from this point of view, would seem to be unsatisfactory and this would mean that the whole system of *"Quantum Physics"* is based on self-referential laws.

Not only that, but, as Kuhn will implicitly confirm in retrospect, every unresolved doubt generates new postulates that will place ever more stringent limitations to protect that originally unsatisfactory structure.

Is it legitimate to ask if is this new formalism that had to adapt to the expressions of nature or, somewhat presumptuously, is the nature that had to adapt to the hypotheses of scientists and their will?

Let it be clear and written in bright letters with large letters: *"Quantum Mechanics"* works and gives comforting results, but everything is based on an unprovable intuition: even if many will surely object that everything works wonderfully.

We can certainly say that the postulates thus expressed are indicative of the inability of a rational, logical and exhaustive explanation, that the theory built on them will always have the weak point of being attacked by sceptics.

One is forced to accept this explanation only because no one has yet found a better one.

What happened with Heisenberg's matrices should be an example: a formalism that produced good results was supplanted by a simpler explanation to use; but which of the two methods was absolutely the best?

Heisenberg's Matrix Mechanics gave a description of how quantum leaps occur by assuming that particles had physical characteristics that could be described by matrices that progress over time.

The Schrödinger Equation, instead of explaining the motion, shows the trend over time of the wave function.

Paradoxically, Schrödinger himself did not like the quantized part of atomic theory and he wrote about it:

> *If we are going to stick to this damned quantum-jumping, then I regret that I ever had anything to do with "Quantum Theory".*

Schrödinger's proposals, however, definitively closed the doors to a romantic conception of physics: from that moment on, the orbits of particles were no longer something determined; it was only possible to imagine electronic clouds around the nucleus in which there was a chance of finding particles. This probability was described by a mathematical function which, in turn, generates other probability functions.

Both the *Theory of Relativity* and *"Quantum Mechanics"* have evolved over time: with the former describing phenomena influenced by velocities and very large bodies, while the latter phenomena in which the objects in play are very small.

At present these two theories are incompatible and therefore we are in the midst of a great crisis that physicists are trying to solve and which will probably lead to the development of a new paradigm.

1927 Heisenberg and his Principle

Heisenberg continued to develop his mathematics and his theories in parallel with Schrödinger. Obviously he didn't love too the mathematical formalism that his colleague had proposed and did not hesitate to show his disappointment that one of his sentences made evident:

> *The more I reflect on the physical portion of Schrödinger's theory, the more disgusting I find it.*

Their view was radically different, even though, after the publication of their works, it was shown that *Matrix Mechanics* and *Wave Mechanics* are mathematically identical.

And Schrödinger had verified this.

But Heisenberg's major contribution came almost at the same time as Schrödinger's work and caused the upheaval of one of the foundations of physics: measurement in an experimental event.

In the world around us, all quantities are measurable in a precise and approximate way to the quantity that you want to measure. So a kitchen scale is fine for measuring recipe components but a goldsmith scale is needed to measure a quantity of gold.

When scientists began to measure quantities in the order of the atomic scale, they immediately realized that the measurements were influenced by the observer because the observer, with his enormous mass compared to the parts involved, necessarily caused many changes that took place in an interaction. .

In 1927, while Heisenberg was on one of his frequent trips to Copenhagen, in a correspondence with Wolfgang Pauli, he announced and described that he had developed a *"principle"* which he himself defined with the word *Ungenauigkeit* that is *"imprecision"* which, for everyone, it will become *The Uncertainty Principle*, according to which, for a particle, the values of velocity and momentum cannot be known at the same time, or vice versa.

The Uncertainty Principle is not limited to just one type of particle or detection methods, but it is intrinsic in nature.

Why are we talking about detection methods?

Because when Heisenberg enunciates this principle he is well aware that in experiments to identify the position of an electron it is necessary to illuminate it with photons which, inevitably, will transmit energy and excite it, causing it to change its momentum and, therefore, accelerate it.

Even using low-energy photons to keep it from accelerating it would still be impossible to determine its position.

The Uncertainty Principle applies to all physical systems because the observer is always an external element that influences, however, the experiment by modifying the results. And if this is irrelevant for large bodies, in the case of the atomic scale the errors have the same scale of the observations and therefore it is highly invasive.

1928 Dirac

Schrödinger and Heisenberg had marked a conjectural turning point: until their times, sub-atomic physics was, on the whole, still understandable. There was a bit of attention to be paid, but the logic that scientists were used was all in all still valid and except for some extreme doubts, everything seemed to be framed: *"Quantum Physics"* had not had a decisive impact.

This was until 1928 when a British physicist with a French name did not appear on the stage of physical research. His new proposal was something so innovative that it would have marked the evolution of *"Quantum Physics"* for a long time. The whole academic world and even his closest friends and colleagues were immediately charmed and Einstein himself, in a letter to his friend Paul Ehrenfest, described him as follows:

> *This balancing on the dizzying path between genius and madness is awful.*

Paul Adrien Maurice Dirac (Bristol, 8-August 1902- Tallahassee, 20-October 1984), was a Nobel British physicist in 1933. In 1925, just twenty-three years old, after Heisenberg, Schrödinger and Born, proposed an even more complete third way and, it would appear, indisputable for the interpretation of *"Quantum Mechanics"*.

Dirac gave *"Quantum Mechanics"* a solid mathematical/conceptual implant that it lacked and that led to the convergence of the matrix and wave vision. This formalism of *"Quantum Mechanics"* is still taught in physics faculties all over the world.

The new equation for the understanding of electron was presented in the 1928.

Schrödinger's equation did not fit to explain the mechanics of particles at relativistic speeds and could only be used when the particles analysed were traveling at much lower speeds than light.

Dirac worked to formulate a single theory that described the behavior of the electron in a universal way and, therefore, proposed an understanding of *"Quantum Physics"* in the light of the theory of relativity that was able to explain the behavior of the electron in any condition and to identify all its characteristics correctly.

His equation also allowed negative energy solutions that will lead to the intuition of *antiparticles* called *positrons* or *anti-electrons* and, to the *antimatter* that Anderson will first detect in 1932's observing cosmic rays.

In the 1928 Dirac also wrote the equation that anticipated the phenomenon of *quantum entanglement* then pronounced by Erwin Schrodinger in the 1935: two microscopic particles that interact for a period of time and with certain modes, if separate will continue to behave like particles, but inexplicably they will have memory of this interaction conditioning each other.

There is a relationship between two particles that come into contact long enough in a closed system that, even if separated at any distance, will continue to be a single system and changes in variables of one affect the other at the same time if they do not come into contact with other external systems. If, on the other hand, the two particles undergo the interaction of other particles, they return to being autonomous entities.

As happened to other illustrious predecessors, Dirac also believed that the result of his work, his equation, was wrong, due to a negative value coming from a square root calculation. Negative solutions were the imprint of *antimatter*, an idea still to be investigated.

The Dirac equation theoretically anticipated antimatter even before it was discovered.

Today the physicist would be a perfect character to play the theoretical scientist: a handsome dark man with a shy character to the point of excess. Not very affable, he did not like to talk either with others or with himself, so much so that in Cambridge some colleagues had ironically coined the *Dirac measure* as a system of magnitude of silence. There are many anecdotes about him: George Gamow, a Russian physicist and Arthur Koestler, a Russian journalist, said that Dirac received an English copy of Dostoevsky's Crime and Punishment from his Russian physicist colleague Peter Kapica. When he asked for his opinion, his only comment was: "Not bad. But in one chapter the author made a mistake. He said that the sun rose twice in the same day ".

Niels Bohr talking to Ernest Rutherford, said about him: *This Dirac, seems to know a lot of physics, but he never says anything.*

His theories were born only from his imagination, without any experiment. When asked to explain his theories of *"Quantum Mechanics"*, Dirac replied that

they cannot be explained in words.

He has studied the mysteries of *"Quantum Physics"* and solved them, mysteriously. His best fans were his colleagues: in fact, despite his shy nature, he managed to be accepted by all the most famous colleagues of his time. And they said about him:

> *Dirac's great discoveries were like exquisitely carved marble statues falling out the sky, one after another...*

Freeman Dyson said of him:

> *He seemed to be able to conjure laws of nature from pure thought.*

His name will remain inextricably linked to the equation he formulated and which was adopted outside the physical world, the only mathematical formula reported in this text:

$$(\partial + m)\psi = 0$$

Surely someone will have noticed it on some graffiti, or tattooed on someone's body: many like to call it the *Equation of Love*, yet it tells of a *quantum reality* that we still can't explain today, *entanglement*. This tangled intertwining between two particles has been associated by many, to the same intertwining, the same disturbance that is established between the hearts of two people who love each other.

Dirac, therefore, is seen as a great romantic physicist.

Dirac's hypothesis on the existence of antimatter stunned the scientific world: it was a too much revolutionary concept to be accepted lightly, at least until 1932, when Carl Anderson, a young physicist at the California Institute of Technology, while carrying out an experiment to understand the nature of cosmic rays, confirmed Dirac's intuition on antimatter.

The passage of particles through a *cloud chamber* [12]highlighted among the many traces, an unexpected one, which Andersen identified as corresponding to the passage of a particle that had the same mass as the electron but with an opposite electric charge, that is positive: it was the first revelation of the existence of the antielectron, which today we call *positron*.

Since then, antimatter research has become more specific and today we are able to generate antiparticles in high-energy laboratories around the world.

Dirac's predictions and Anderson's confirmations forced us to revise the belief that subatomic particles were immutable: *electrons and positrons* can generate or can cancel each other out revealing other forms of energy.

Since then, many investigations in subatomic physics have been directed towards the discovery of new types of particles that in many cases exist only for a very limited time.

---

[12] The cloud chamber was first used at the University of Berkeley in 1938, and allows the detection of particles. For its construction, very simple and easily achievable even at home, evaporated alcohol is used which originates a "cloud" which can be very influenced by elementary particles in transit, and more specifically by some types of cosmic rays.

October 29, 1927: Fifth Solvay Conference and Copenhagen Interpretation

Swedish Alfred Nobel (Stockholm, 21st October 1833-Sanremo, 10th December 1896) and Belgian Ernest Solvay (Rebecq, 16th April 1838-Ixelles, 26th May 1922) were two chemists who capitalised their knowledge by setting up two industries that made them enormously rich. They were also two friends who wanted to return to humanity the well-being received with initiatives that promoted development and scientific progress. Nobel perfected the use of nitroglycerin as an explosive and invented dynamite: these two discoveries made it enormously rich. His legacy allowed him to establish a foundation that established a Nobel Prize to reward in the years to follow the most distinguished minds in their respective fields of membership of the human race.

Solvay, a brilliant and inventive character, accidentally created sodium bicarbonate and optimized its production and distribution worldwide. In order to distinguish itself from the Nobel Prize initiative, Solvay decided to organize a week-long congress every three years in Brussels to facilitate the comparison, exchange of point views and innovative ideas, bringing together in a stimulating and brilliant context twenty of the best minds of the period, in physics and chemistry.

From 1911 till today, the Solvay Congresses are still organized and considered important meeting places for the development of the progress of physics and science in general.

Certainly, what took place in October 1927, was worthy of special mention, both for the human capital that participated in it, and for the events that marked the week of its development, worthy alone of being a topic to be dealt in a separate book.

Let us imagine being there and, let us imagining the context: 29 participants, 17 Nobel Prize winners among those who had already achieved it or who would have achieved it and only one exceptional woman, Madame Curie, with two Nobel Prizes and the fundamental contribution to the discovery of radioactivity. Then everyone else who had contributed in some way to the birth and perfection of atomic physics. There were also the Founding Fathers of *"Quantum Physics"* and the future implementers of its *Mechanics*.

The theme of the congress was *Electrons et Photons* and was predicted to be charged with tension even before it began, much more like a boxing ring than at a conference of exceptional minds.

Einstein, strong of his finally established ideas and therefore universally accepted, did not harm his colleagues. For example, referring to Max Planck, he said, regarding the short period spent together for Congress:

> *[Max Planck] was one of the finest people I have ever known... but he really didn't understand physics, [because] during the eclipse of 1919 he stayed up all night to see if it would confirm the bending of light by the gravitational field. If he had really understood [general relativity], he would have gone to bed the way I did*

Back Line: *Auguste Piccard, Émile Henriot, Paul Ehrenfest, Édouard Herzen, Théophile de Donder, Erwin Schrödinger, JE Verschaffelt, Wolfgang Pauli, Werner Heisenberg, Ralph Fowler, Léon Brillouin*

Half Line: *Peter Debye, Martin Knudsen, William Lawrence Bragg, Hendrik Anthony Kramers, Paul Dirac, Arthur Compton, Louis de Broglie, Max Born, Niels Bohr.*

Front Line: *Irving Langmuir, Max Planck, Marie Curie, Hendrik Lorentz, Albert Einstein, Paul Langevin, Charles-Eugène Guye, Charles Thomson Rees Wilson, Owen Richardson.*

Everyone wanted to see the intellectual "clash", which surely would have been on a minefield as *"Quantum Physics"* was at that time, between Einstein, father of the *Theory of Relativity* and Bohr, creator of the so-called *Copenhagen Interpretation of Mechanics Quantum.*

Niels Bohr and Werner Heisenberg, who had contributed to the evolution of *"Quantum Mechanics"*, in a context of mutual growth of knowledge, collected and expressed opinions on its meaning and accepted open and informal comparisons with other physicists.

Their method, which today we would call *brainstorming*, proved to be so valid in providing different interpretations, that it still constitutes an excellent methodology for tackling such complex topics in university physics faculties.

The heated comparison that ensued, allowed to express a synthesis of the works, until then carried out in a sparse manner, hypothesizing that the wave function of a particle described a set of possibilities, all simultaneously present, until the process of measurement, carried out by an external observer with any instrument, forces the system to collapse into one and only observable state.

Therefore the same quantum particle is wave until the moment of observation, particle at the time of measurement.

The Copenhagen Interpretation of *"Quantum Mechanics"* offered the great advantage of not affirming details directly on physically unobservable quantities and with few prerequisites and, it allowed proposing a conceptual system that well described the experimental facts available at that time.

In the second half of 1927 the heavy differences of opinion that arose above all between a small group of pioneers of *"Quantum Physics"*, centered on the Copenhagen Institute, with Niels Bohr and Heisenberg in the lead and subsequently with the Austrian physicist Wolfgang Pauli, managed to be smoothed out afterwards. the exposition of Bohr's concept of complementarity, whereby any physical phenomenon can be studied in two different and *complementary* ways that depend on the modalities of the experiment set up to study it.

In this way, the light, for example, could have a double manifestation: sometimes a wave and sometimes a particle, according to the mode of observation. Although apparently in contrast, only the double aspect provides a complete description of the phenomenon.

Brian Greene in his *The Hidden Reality* provides a very clear interpretation of this way of conceiving *"Quantum Mechanics"* and its experiments:

> *The standard approach to "Quantum Mechanics", developed by Bohr and his group, and called the Copenhagen interpretation in their honor, envisions that whenever you try to see a probability wave, the very act of observation thwarts your attempt.*

Bohr, after proposing his results at a conference in Como, Italy, in the summer of 1927, proposed them again at the Solvay Conference in what became the official formulation of the Copenhagen Interpretation. Due to its inherent ambiguity it became, however, casus belli for the heated discussions between Bohr and Albert Einstein.

Seven days of confrontation between two people who respected each other but who had different visions: the physics, grateful, thanked.

Surely Bohr, with his experience and his studies dedicated to atomic structure and *"Quantum Mechanics"*, was more specialized on the issues under consideration, but Einstein, by now strong in the great popularity of his theories, had a greater influence and had become a point of reference for any subject that concerned science, physics and other aspects of life as well.

Einstein Bohr Duel

We can imagine Bohr as a "large man": tall, sturdy and always smiling. Passionate pipe smoker always had a dandy air with the ubiquitous pipe between his teeth but, contrary to appearances, he was always ready to argue sharply and overturn conventional ways of thinking and theories.

Einstein had great respect for him and spoke of him enthusiastically, calling him a genius of physics. In addition to a purely material view of the atomic world, he had also drawn philosophical conclusions from the physics he was studying.

In a few decades, his works had changed all the knowledge acquired by humanity until then, supplanting them with innovative theories that are difficult to understand.

Einstein, on the other hand, needed no introduction: he was the one who had come up with a strongly alternative theory out of thin air.

For everyone he was a genius, pragmatic and realist: these two opposing ways of understanding physics and the lack of a neutral reading of the Copenhagen interpretation made it impossible to define an objective validity also due to the lack of directly observable experimental data.

This lack of objectivity has continued to the present day, since now there is no evidence produced by the experiments, but only intuitions that justify the results of the measures hypothesized by the Copenhagen interpretation, structured on three points:

- *The Heisenberg Uncertainty Principle* of 1927,

- the assertions of Max Born in 1926, which annulled the *principle of causality*: the results were irreparably influenced by the instrument of observation, so that the same observation could never provide a certain result. The result was also meant only as the probability of a result in a given state, in accordance with the Schrödinger wave function

- the concept of *"complementarity"*: two aspects of the same description which are mutually exclusive but which are both necessary to understand the phenomenon.

Almost every day the two scientists confronted each other, especially on the *Uncertainty Principle*, continuing the dispute even during official lunches. At breakfast Einstein subjected Bohr the night before studied experiments that could infect the *Uncertainty Principle* and Bohr sought answers that would satisfy the interlocutor. But the next morning, another question and another day looking for answers.

One of the most ingenious mental experiments Einstein proposed to refute Heisenberg's principle was the one in which he put energy and time in relation: he succeeded in demonstrating how it is possible to obtain a measure of both the energy attributable to a particle and the time in which it was emitted.

The experiment was to imagine that a light beam would come out of a box at a precise moment. The weight of the box that we measure before and after, applying the equation of the small relativity $E=mc^2$ that relates energy to mass, will give us the value of the energy of the light beam; otherwise the *Uncertainty*

Principle would no longer be valid in this case as the energy of the light beam and the exact moment it came out are known precisely and unlikely.

Bohr was astonished: an experiment so simple but so ingenious seemed to leave no room for a possible solution. But the next morning Bohr presented himself with a solution that took advantage of Einstein's theory of relativity: because the force of gravity influenced, in the act of weighing the box, even the passage of time, it would not have been possible to determine the measure of the precise moment at which the particle departed from the box. And Bohr won this duel.

It must be clear, however, that, despite rivalries, Einstein and Bohr held each other in esteem and contention happened out of sheer thirst for knowledge, respecting their own ideas and opinions: they both sought the truth.

Einstein, like other scientists, physicists, philosophers, did not accept Copenhagen's interpretation of *"Quantum Physics"*, either because it was not deterministic or because it introduced an indefinite measurement method that transformed probabilistic functions into reliable solutions. For him, the physical reality provided by *"Quantum Mechanics"* was incomplete, so he could not abandon the search for an alternative theory and more adherent to a deterministic reality.

Einstein once received a letter from Born, to which he replied:

God does not play dice with the universe!

When Bohr found out, he broke in by inviting Einstein to

Einstein, stop telling God what to do.

In another conversation with Abraham Pais, Einstein exclaimed:

The moon exists only when I look at it?

always referring to the probability of a manifestation rather than its certainty.

The clash lasted their entire life, with ideal experiments imagined by Einstein to contradict the quantum view, often with difficulty countered by Bohr and the other great physicists of the Copenhagen school convinced of the goodness of their probabilistic interpretation of the physics of the atomic world.

Einstein remained convinced until his death of the objective existence of a real world independent of the presence of an observer. He was in fact convinced that the paradoxes of *"Quantum Physics"* are attributable more to an intrinsic weakness of incomplete theories than to the whims of a nature irreconcilable with the capabilities of a human mind.

For Einstein such position would have led to losing sight of the very criterion of reality and the reassuring materiality of the objects of the world and until the last moment of his life he continued to search for the *Theory of Everything (TOE)* that would have allowed to formulate a single theory that would have including all other theories in a more complete view.

Bohr never gave up on his dualistic interpretation of the subatomic world and when in 1947 he received the "Order of the Elephant" from the king of Denmark, one of the most important honors of that country, finding himself in the need to find a coat of arms that represented him, adopted the symbol of Yin and Yang, completed with his motto: *contraria sunt complementa*.

He wanted to be the "summa" of his worldview.

Bohr died in 1962, seven years after Einstein, in total intellectual solitude: in those long years he had lacked the only interlocutor who understood him, so much so that he was often seen discussing alone, as if he were talking to Einstein in his presence.

Wave Function Collapse

We have seen that wave function, in *"Quantum Mechanics"* describes the behavior of subatomic particles, usually electrons, in the shape similar to waves.

What is a function? A function is a mathematical mechanism that transforms something into something else according to predetermined laws: so we can imagine the function as a mechanism with two or more outputs. What you insert at the entrance is manipulated to give something different to what was introduced, just as if it is a kind of machine where if you enter a number, it is processed and produces another result other than the number introduced.

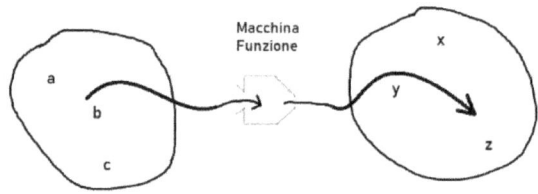

The wave function, therefore, is a mathematical tool that inserted known parameters produces, as a result of manipulation, an equation that describes a wave.

A mathematician or a physicist, looking at the result of the final product, can understand what it refers to and recognize it as a mathematical structure that identifies a wave.

The parameters we enter are those we measured during an observation; or, having a generic wave function, we can insert parameters that we imagine to be consistent and verify that they belong to something familiar, in this case a wave.

Because the physicists in their hearts are also great nostalgic, they wanted to call this function the ancient Greek letter Ψ, which reads *Psi*, used also as a symbol to indicate psychology. There was a remarkable sense of self-irony behind all of this because having something in his hands that made him lose sleep, the wave function, it was called with the name of an inexpressible desire, that of the knowledge of our psychology. Thus Ψ became the symbol and the name that in mathematical and physical formalism usually indicates a wave function.

In 1926 a very distinguished gentleman named Erwin Schrödinger, let's remember him because he'll also be dealing with cats that were probably infesting his office, in a normal day went at his office in university and, like every day, he began to work on an equation that he loved to study and day after day, he was able, by touching a little here, a little there, to elaborate until he produced a very powerful and performing mathematical engine. This allowed him, by entering at the entrance the mathematical parameters of the behavior of the quantum particles, such as mass, spin, electric charge, etc. to generate one or more functions in output. The functions resulting from the original equation, obtained by processing the parameters of the quantum particles, had the mathematical form of a function describing a wave.

The resulting function was another machine, which also transformed something into something else: Schrödinger had used a mathematical machine to build another mathematical machine.

The result, the final machine, was another function, called *wave function* and, the equation Erwin worked on every day in his honour, was called *Schrödinger Equation*.

Let us give a practical example of the operation of the engine that generates the wave function.

We managed to lock an electron in a box, trapping it in magnetic fields. The electron, as we have seen in previous chapters, emits a wave function. Analysing and calculating it will tell us the probability that the electron inside the box is in a certain position. Inside the box we have six magnets, all with the same intensity, that generate six magnetic fields in six different positions inside the box. By entering into the wave function the value of the forces of the magnetic fields, along with other related factors, such as mass, spin, electric charge, etc., and processing the calculations, the final result will be a probability function that will give us the possible position of the electron in the box. The result of the calculation is a set of probabilities where we could find the electron because *"Quantum Mechanics"* is probabilistic.

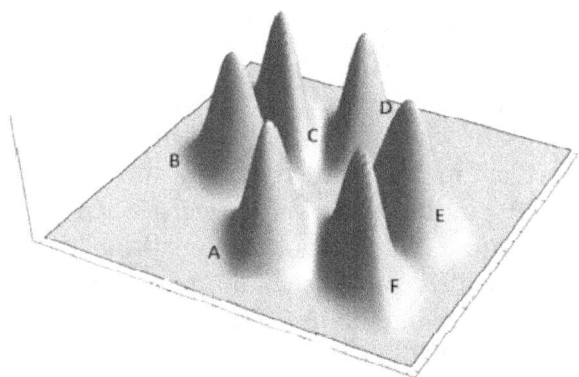

Unknow Origin Design

Thus, if instead of the electron we had any subatomic particle, we could use the wave function to calculate, predict and verify the behaviour of subatomic particles in experiments, as is the case, for example, in Geneva at the LHC accelerator.

Normally this method is used to describe the behaviour of electrons because it is not as precise with particles moving faster.

Paul Dirac subsequently further upgraded the Schrodinger equation engine, just as is the case for Daytona racing cars engines that are developed race after race to make them more efficient, so that they can also be used with fast-moving quantum particles, I mean, all those particles that are connected to *Einstein's Special Relativity*, like photons.

Nowadays, the original wave functions have become much more complete thanks to the various processes they have undergone, a bit like a 30-year gasoline engine that has gradually become a Daytona racing cars turbine engine, thanks to the work of the many mechanics who, over the years, have invested time and energy in it and hey take the form of *Quantum Electrodynamics (QED)*, which is used to calculate the behavior of slow and fast particles.

In fact, not all physicists agreed or agree this approach; in fact the physical meaning of the wave function is a reason for eternal debate among the quantum physicists who ask themselves whether the wave function really described a real physical wave.

Problems with these engines arise when an experimenter wants to measure the parameters of a quantum particle, i.e. when he wants to observe an event.

Because of the subatomic dimensions of the objects to be measured, the same experimenter, become integral part of the experiment, disturbing the results. In simpler terms, the measurements obtained are affected by the presence of the observer. It is as if the same moment we decide to make the measurement we immediately stop the whole experiment: this would allow us to see exactly what happens at that precise moment to the particle we are studying because it stops. This moment is identified by the term *Wave Function Collapse*, an expression we use in the study of *"Quantum Mechanics"* and in the definition of *Copenhagen Interpretation*. In other interpretations it may have different meanings.

Thus, in accordance with the double nature, wave and particle, the collapse of the wave function is the transformation from a wave function, which describes the probability of finding a particle in a given place, to a localized particle with certain parameters identified by numbers.

By definition, the wave function cannot be used to calculate how probabilities change when particles are detected in an experiment.

Let us try to understand this by way of an example: suppose we prepare a pudding with a cooking robot. We insert all the ingredients and get the dough. We will know that in the dough, among other ingredients, we have also put apples pieces: we could say that it is an apple pudding, but we will never really know where each piece of apple has ended up. In theory, we could take a bite and not find a single piece of apple. We could then say that we ate a pudding but we could not say that we ate an apple cake.

This conclusion, applied to particle detection experiments, tells us that in the wave equation there are different possibilities of actually finding a particle at a given time but we are not at all sure: it is mathematics that affirms it without the observation actually confirming it. So the result of the wave collapse is as if it were detected by the eye rather than detected by an actual observation.

These are certainly conclusions that leave us startled and bewildered. It seems that the calculation of behavior is not precise and that we never really know where the particle really is. If it's any consolation, so it is. This reasoning is correct and it is equally disconcerting for physicists that, as such, having fully understood a phenomenon, they should be able to predict its behaviour. So, at least, it was in classical physics.

Yet, even repeating the experiment hundreds, thousands of times, the result is always the same and the wave function provides systematically correct information, and always predicts exactly what are the chances that electrons are in a specific positions.

Recalling the previous example, it is as if, knowing the number of pieces of apple that we have cut, we could calculate the number of pieces that we would find at each bite and, if we did the test, we could check each time that we have reached the right value.

Because of this apparent but unexplained contradiction, many physicists use wave function without fully understanding the physical reality it describes.

This type of approach according defined by Feynmann as the method *Shut up and calculate*, precisely because to the questions of his students on the contradictions of the mathematical method, he used to answer in this way.

"Quantum Physics" Comes of Age

The universe is full of magical things patiently waiting for our wits to grow sharper.

Eden Phillpotts

1929 First Particle Accelerator

Ernest Orlando Lawrence (Canton, August 8, 1901 - Palo Alto, August 27, 1958), was an American physicist Nobel Prize winner in 1939. He conceived the cyclotron in 1929 while he was at the University of California, Berkeley. As its name indicates, it was a circular machine in which beams of electrically charged particles, such as electrons and atomic nuclei, were collided after being accelerated to very high speeds. Two D-shaped electromagnets directed the charged particles in semicircular paths; in the small space between the electromagnets an electric field of very high intensity was applied which accelerated the particles, which began to spin at ever higher speeds, until reaching the desired limit. These accelerators became the primary tools for nuclear and particle physics research of the twentieth century.

Later, all the machines that studied most subatomic particles relied on Lawrence's original cyclotron. One of these, very similar to the original project, was used during the design of the first atomic bomb to separate uranium-235.

The first cyclotron designed by Lawrence was made by an American physics student, Milton Stanley Livingston (1905-1986): it had a diameter of a few meters and used thirteen thousand volts.

Current particle accelerators such as the LHC in Geneva, the largest and most powerful to date, has a diameter of about 8 km, and it is placed in a tunnel almost 100 m below the ground and uses an energy of almost 14 teraelettronvolt, a huge number.

Roaring Twenties

Early years of twentieth century produced epochal political economic changes in societies that became a driving force for all these new discoveries in the physical field. Everything contributed to push humanity to a real leap forward.

In United States they were called the *Roaring Twenties*, It was a way of saying to indicate a formidable period of economic growth in which the consumer boom, the progressive lifestyles, at least in some areas of this large country, and the changes of the social organization gave the feeling of entering a new age of gold.

In those years the Americans who lived in the cities exceeded the number of those who lived in rural areas. Between 1920 and 1929 the economic growth led most Americans into an unknown and rich "consumer society".

For the first time there was a new social phenomenon, a massification that led many to radically change their lifestyles: they bought the same objects, promoted through the new advertising industry, first on paper and then on spots, the advertising industry was born, the first chains of shops appeared and everyone danced listening to the same music. At least for the young people of the big urban areas, the '20s were really roaring.

In those years the United States converted from a war economy, the "Great War" ended in the 1918, into a peacetime economy. In this decade a period of prosperity began and the foundations were laid for transforming America into the richest nation on Earth thanks to the culture of consumerism.

Henry Ford sold more than fifteen million T models in 1927, an enormous number that could be achieved thanks to the introduction of assembly chains in the factory: things that had never been seen before. The car industry has led to the development of roads network made up of paved roads and motorways.

The radio became a ubiquitous domestic appliance in every home, at least in America.

In 1926, the invention of Technicolor came after the first sound movie was launched in the 1922: the movie industry became a driving force for the entire American industrial complex.

Charles A. Lindbergh, using the Spirit of St. Louis, in 1927 crossed Atlantic Ocean becoming the precursor of the future aviation industry.

In 1923 the first mass vaccination against diphtheria was made and in 1929 Alexander Fleming discovered penicillin randomly, studying a mould on a colony of staphylococci.

USA were just like wonderful places were investing more in social economic well-being and technologies than in science studies.

At the same time, Old Europe faced its history in a completely different way.

The First World War, both among winners and among losers had left very heavy straits: many families had no longer those men who with their arms works would guarantee an income and the poverty of means of subsistence pushed a large number of Italians, Poles and Irish to emigrate to the Great American Dream.

Especially in Germany, because of the sanctions imposed, the economy was brought down by a hyperinflation that forced the productive system to fail, increasing unemployment and leading to poverty for many families. So in 1923, also the result of the social rebellion, in Munich, Adolf Hitler tried for the first time to take power, which unfortunately he will achieve ten years later.

The Twenty Years if on the one hand showed a society benefiting from a general explosion resulting from the human potentials that were rooted in the Industrial Revolution and in the expansion of global markets, on the other hand, at the same time, paid the consequences of short choices, also dictated by nationalism that everywhere, even in America, laids the foundations for the Great Depression of 1929.

In 1922 Benito Mussolini came to the government in Italy, with the promise of a better future.

India chose independence with Gandhi's non-violence policy.

It was in those years that one of the most serious crises still afflicts us is sinking into the ground: the division of the Middle East and the origin of the Palestinian issue.

Some European citizens looked at America and let themselves be dazzled by fighting their difficulties: so many people in Moscow, Berlin, Paris became the new beacons of fun, of beautiful life, of luxury and of culture.

In this historical and social context, sciences had the greatest discoveries of *"Quantum Physics"*.

It would seem that the most idealistic and pure scientific world was locked in a bubble filled with the brakes of those years. Scientists were a world apart, little inclined to mass popularity.

The times were changing and pure and unconditional science gave way to a science that continued to progress because it was functional to the new war industry that was preparing the war secretly.

After 1930, the history of atomic physics and his protagonists became very fragmented: many scientists perceived winds of persecution blowing, left Europe receiving hospitality in the United States, that in this way they suddenly found such a scientific capital that soon made them the new point of reference of almost all sciences.

Scientists around the world were no longer able to develop their work organically and physics, suffered from it, began to proceed in a diffuse way.

Physics Goes to War

We turned back to look at Hiroshima. The city was hidden by that awful cloud... boiling up, mushrooming, terrible and incredibly tall.

No one spoke for a moment; then everyone was talking.

I remember (copilot Robert) Lewis pounding my shoulder, saying 'Look at that! Look at that! Look at that!' (Bombardier) Tom Ferebee wondered about whether radioactivity would make us all sterile.

Lewis said he could taste atomic fission. He said it tasted like lead..

Paul Tibbets

Heisenberg Cold Case

Scientists became tools for war machine and, in spite of them, had to confront the harsh political reality, often suffering heavy political conditions that forced them to make field choices, abandoning the paths taken and severing on many occasions' contacts and friendships that allowed the exchange of new experiences.

In many cases friends and colleagues of the past turned into opponents, if not persecutors.

In 1929 Heisenberg and Pauli presented a joint paper about *"Quantum Theory" of relativistic fields*. In 1932 Heisenberg received the Nobel Prize for Physics, and was officially recognized as father of *"Quantum Mechanics"*. This prize, in his opinion, should have been shared with Max Born and Pascual Jordan, his collaborators on matrix mechanics.

Since that moment and certainly against his expectations, Heisenberg's existence became troubled because he had to deal with the history of his country and with the acceleration imposed by the military forces to discover new weapons: Nazism and the race for the atomic bomb broke into everyone's lives.

Philipp Eduard Anton von Lenard (7th June 1862 to 20May 1947) was a German physicist Nobel Prize winner in 1905. On the wave of emerging patriotism and nationalism, wrote in the 1914 the book *England und Deutschland zur Zeit des grossen Krieges*, in which he accused the British scientists to systematically copy the work of German scientists. Few years later, when Adolf Hitler appeared on the political scene, the physicist immediately embraced the National Socialist ideas, earning his esteem and, on the 15th of May 1926, in a party meeting in Heilbronn, met him personally. He became one of the Fuhrer's trusted scientists and embraced his anti-Semitic ideas, which, unfortunately, were also creeping into the minds of such illustrious scientists.

In 1933 Adolf Hitler became Chancellor of Germany and many German theoretical physicists, especially those most oriented to research on *"Quantum Physics"*, including Sommerfeld, Planck and Heisenberg himself, found themselves hampered and marginalized, especially by Philipp Lenard and Johannes Stark, which were particularly listened from the new political power group.

Arnold Sommerfeld openly supported Einstein, who had Jewish descent and, as far as possible, stood up against the groups that supported anti-Semitism, but he was alone and also his colleagues, including Planck and Wien, did not have the same momentum as he had to oppose the most aggressive members of the Nazi Party: he and other scientists, including Heisenberg, who did not conform to this state of affairs and refused to accept the laws on racial discrimination, were defined by the nickname of "white Jew," because they supported Jewish science despite their Aryan origins.

In 1935 Sommerfeld was forced by events to leave the theoretical physics university chair in Munich and this event became the determining cause for which will be remembered as the *Heisenberg affair*.

In the first place, among the possible replacements, Heisenberg's name emerged, who, despite the Nobel Prize and all the best references, was initially hampered and then involved in a dispute that put his own life in danger.

The physicists liked by the politic bosses gathered in the Deutsche Physik movement, against those scientists especially Jews who advocated theoretical physics, did not appreciate Heisenberg.

In 1936 Lenard published his racist and political ideas against other scientists, who did not openly support National Socialism, publishing four texts, titled *Deutsche Physik*, in which he reasserted his convictions on the need for a pragmatic *German physics* against dogmatic *Jewish physics*.

Deutsche Physik became a clearly anti-Semitic stream of thoughts against *theoretical physics*, *"Quantum Mechanics"* and *The Theory of Relativity* and, of course, especially Jewish proponents of these theories.

In 1938, the Schutzstaffel and their leader Heinrich Himmler also began investigating Heisenberg, who replied with an editorial to dispel any suspicion from him.

On the 21st of December 1938 Otto Hahn and Fritz Strassmann, in a Germany where the winds of war were blowing ever stronger, for the first time in the history of mankind divided the first atom of uranium: scientists and above all physicists around the world, immediately realized that they could control almost unlimited energy and also have the possibility to build a potential weapon of resolution with devastating effects.

Fortunately many had already moved away from Nazi Europe, finding shelter in Britain or in United States.

Nuclear physicists emigrated to the United States, including Leo Szilard, Albert Einstein, Enrico Fermi and Emilio Segre, who understood the danger on the horizon, worried about the effort the Germans were making in building an atomic bomb, They tried in all possible ways to keep President Roosevelt and the high levels of command informed of the danger that was going on, given also the disadvantage of a couple of years accumulated against German scientists.

Roosevelt decided to set up the *Uranium Committee*, a committee of military and scientific experts to study the feasibility of a nuclear reaction.

On April 24th 1939 Paul Harteck, a chemist and physicist from Hamburg, and Wilhelm Groth, his assistant, wrote a letter to Erich Schumann, head of the German Army Ordnance's weapons research office in Berlin, Heereswaffenamt. The letter described possible military applications of the newly discovered uranium fission technology. Schumann, who was also the grandson of the famous composer, being a general but also a physicist, had immediately understood the potential of this discovery. Unfortunately for him, but for our good fortune, he was completely ignored by the nuclear physicists who were part of the army's experts, and so there was a guilty delay on the part of those who would soon have been co-opted to arrive at the design of a nuclear bomb as soon as possible.

In 1939 Heisenberg returned to Germany from a trip to the United States, where he had visited the University of Michigan where he received an invitation to emigrate, which he refused. Upon his return he became one of the most important reference points of the German plan for the development of military applications in the field of nuclear energy, known as the *Uranium Club*.

In 1941, in a Copenhagen city occupied by the Reich, in tour for some conferences, the scientist also met Bohr: we will never know what they talked about, but almost certainly also of a possible military development of *"Quantum Physics"*.

The Reich pressed him to accelerate, deepening a greater commitment in atomic research to develop as quickly as possible a weapon that could become decisive in the conflict, but Heisenberg did not seem to agree.

In United States, however, due to the slowness of the chain of command, no decision was taken on the development of a nuclear weapon, as scientists had suggested.

Until, in the spring of 1941, the similar British organization called *MAUD* of the *U.S. Uranium Committee*, thanks also to the reports of the agents on German territory, confirmed the feasibility by German scientists of a weapon with a high destructive capacity based on the atom and called for cooperation with the United States.

The U.S. government initially entrusted atomic research to the *S-1 Committe*, except to hand it over to the army when it realised the considerable difficulties.

And so, while allied armies in Europe were fighting a Germany that was now torn apart by the battle, two groups of the best nuclear scientists in the world were desperately at work. A team in the desert of New Mexico, reorganized in the most secret *Manhattan Project*, was eagerly trying to assemble the atomic bombs that would be used later in summer; the other group constitued of highly trained and brilliant physicists and technicians, fought in southern Germany to build a nuclear reactor to be delivered to the Reich.

Everything concerning the atom was secreted by both groups that had dedicated the best minds available at that time to solve the problem. The Germans were able to recruit formidable scientists, including Heisenberg, Otto Hahn, Max von Laue and Carl von Weizacker, three of whom had won Nobel Prizes for physics or chemistry. They were assembled in one place, with adequate resources and were supposed to produce the ultimate weapon for the Fuhrer.

It was therefore crucial that US intelligence officials knew as much as possible about Germany's progress on nuclear weapons.

In the 1943's, because of the hammering bombings on Berlin, German scientists had great difficulty in carrying out any work. It was therefore necessary to find a new place, possibly outside the war zone, where they could continue their work undisturbed. The project was then moved to southwestern Germany in a narrow valley along the river Eyach, in an area difficult to reach by airplane bombers and by possible invasion, especially of the Russians.

In the basement of a local church the rock below was excavated to build a small reactor.

At the same time, in United States, Robert Oppenheimer and his team of scientists worked with the same goals at the Manhattan Project, well aware that there was a two-against-time race going on.

General Leslie Groves, head of the Manhattan Project, devised a plan to slow down the work of the Germans before they could achieve their goal: it was necessary to organize a mission to track down the German group, destroy the facilities and any uranium they had and, if possible, captures the scientists working on the project snatching any paper material that was traced.

The mission was called *Alsos Mission*, Greek word for woodland, because German scientists had their headquarters in a forest. Operations were controlled directly by Project Manhattan personnel. On the field, under Colonel Boris T. Pash and Dr.

Samuel A. Goudsmit, there was a task force comprising seven military officers and thirty three scientists.

Pash was the military intelligence officer, head of security for the Manhattan Project and military commander of the mission. Goudsmit was a Dutch-American physicist and archaeologist and ran the mission's scientific team.

In April of 1945, the original plan for Alsos's task force foresaw to join the American army in its attempt to cut the roads in south of Stuttgart. The team would then head to Hechingen with infantry men detached from the Corps.

General Alexander Patch, commander of the 7th U.S. Army, with an armed division and three infantry divisions, was heading for Stuttgart, coordinating with the First French Army of General de Lattre approaching through the Black Forest. United States would have taken control of Stuttgart, being in the American occupation zone. But Charles de Gaulle, on behalf of the French Grandeur, pressed de Lattre to get to the main German city of the region, Stuttgart, so that it could be conquered by the French army.

The French, who were unaware of *Alsos Operation*, were threatening to blow up the whole operation.

Eisenhower did not want the French to arrive sooner because under no circumstances German scientists and their research data had to fall into their hands. The people responsible for the Manhattan project and the entire Army Staff distrusted French scientists, openly sympathetic to Russian communists and, above all, believed that Madame Joliot-Curie, the leading French scientist, was herself, directly, a communist spy who would hand over the captured German material to the Russians, who would find such valuable material without harming.

The results of the atomic bomb research had to be Anglo-American at all costs.

On April 19th, the French forces conquered Stuttgart.

The Anglo-American major state, in panic, authorized a handshake: he ordered a body of American Army to cut the lines of French operations to take and hold Hechingen, regardless of the French will.

So the team of allied scientists, with the support of the combat troops, moved without French permission to reach the German scientists in Hechingen.

Pash, who was a bold and ingenious soldier, made his way bluffing with French allies: as he crossed a bridge that gave access to the roads leading to the place where the German scientists were, when together with his team he was stopped by the French soldiers commanded by General de Lattre, he also made a long speech with the help of the interpreter who translated everything very slowly.

While this interview was going on, the rest of the column snuck across the bridge and, before the French realized they had been deceived, Alsos's team and its supplies disappeared along the way.

Using other subterfuge, finally, on the morning of April 24th, Colonel Pash and his men arrived in Hechingen, which they conquered after a firefight with light weapons.

What they found was only the construction of a test reactor and not a bomb project.

The Alsos team quickly tracked down all the research documentation and, most importantly, the German researchers. They captured Hahn, von Weizacker, von Laue, and tracked the others to their offices and homes.

Pash and Goudsmit personally questioned the scientists, who were happy to surrender to American soldiers.

The American soldiers dismantled the structures by loading everything they could on the trucks, along with uranium found. The partially completed reactor was hidden in the church basement.

The Germans had not been able to complete the project just because they had not been supplied with additional quantities of heavy water and uranium rods in time.

The soldiers instead of blowing up the reactor with the consequent destruction of the baroque church limiting them to targeted destruction that saved the building from the ground.

German nuclear scientists were let out of Hechingen under the French nose and in a month arrived in England. Escaped to shooting they were interned at the secret Farm Hall facility near Cambridge where they were placed under close surveillance. Unbeknownst to them all their conversations were recorded by microphones installed everywhere in the buildings of the residence. This operation, known as *Epsilon Operation*, allowed agents of British secret service to listen and record the speeches of German scientists in full.

In 1993, the whole arrangement was discredited by military secrecy and allowed all fans to know what happened...

The physicists were mostly talking about their failed program and the latest events that led to the destruction of Hiroshima: German scientists were surprised how quickly the Americans had made and used the bomb they did not believe to be transportable.

Some were openly against Hitler's National Socialism which they detested, but others were still loyal to him.

The mystery remains as to why, having started with such a great advantage, the Germans were not able to materialize it: someone, including Heisenberg[13] himself, had deliberately slowed down the researches because they became aware of the follies of the Nazis?

In May 1945 the Second World War ended. Heisenberg was released in 1946 and settled in Göttingen working as director of the Max Planck Institute for Physics until 1958, then of the Max Planck Institute for Physics and Astrophysics in Berlin until 1970. He continued to lecture around the world and published articles, also interested in superconductivity, cosmic rays, as well as participating in councils, commissions and associations.

Lenard certainly adversely affected German scientific progress due to his strong opposition to collaborating in particular with Albert Einstein. This diatribe was the subject of the book "The man who persecuted Einstein: how the Nazi scientist Philipp Lenard changed the course of history" by Bruce J. Hillman, Birgit Ertl-Wagner and Bernd C. Wagner.

---

[13] It is interesting to know that Werner Heisenberg, director of the German nuclear team, was good friends both with Heinrich Himmler, Reichsfuhrer of the SS and the Gestapo, and with J. Robert Oppenheimer, the physicist in charge of the Manhattan Project.

The Neutron

In 1932 James Chadwick discovered that in the nucleus of the atom, in addition to the proton and electron, there was also another particle: the neutron, which was devoid of charge and mass.

1930 Gödel

In New Jersey in the United States there is a small town surrounded by greenery, very quiet, where is located the fourth oldest university in the United States: Princeton University.

Princeton is a prestigious private, medium-sized elite university with its 5.321 students. Access to Princeton is still a great privilege, and students completing their studies up to graduation are guaranteed to find a job almost immediately with immediately very high salaries.

Along its large tree-lined avenues, which hide behind the ancient trunks and majestic houses, it is easy to meet students who walk and chat with each other.

Imagine taking a leap in time, in an indefinite period after the end of the Second World War and walking along one of these avenues: you can hear the far blows of the balls from the players of a golf course and in the distance two men, lonely, come to meet us. Coming within voice range, we would hear them talking quietly in German.

At that time it was easy to hear different European languages spoken among the locals: in 1930 the IAS, Institute for Advanced Study, was founded thanks to a rich donation by Louis Bamberger and Caroline Bamberger Fuld, who, following the suggestion of their friend Abraham Flexner, a well-known theoretician of education, invested their money in the service of more abstract research.

The idea was to have a flexible laboratory institute, which would become a refuge and research place for students and scientists who could thus devote themselves to the study only without being pressured by the immediacy of the results.

A simple, comfortable, quiet place without being enclosed, but remote enough to maintain the necessary confidentiality; no problem had to harass its visitors who did not have to endure any pressure. An ideal place to allow students an impartial and free research from any ideological conditioning and from fears of expressing concepts other than those normally used.

Its university attendants, still today, have complete intellectual independence and are completely free from administrative responsibilities or concerns.

This Institute, which seems to be completely outside of any convention and constraint, had two fundamental principles: the first demanded that all those who would become associates should be accepted and evaluated only on the basis of their abilities, without any race, creed or gender constraint. The second is that the Institute should enable them to carry out independent and non-profit-making research, thus pursuing only the attainment of knowledge.

Among the guests were well-known figures among the academics of that time: Einstein and Von Neumann, Gödel, Oppenheimer and others to their level.

In Vienna the genius of Gödel had not received the esteem and recognition he deserved, also because of the now prevailing national socialism. At Princeton, instead, which had now become the reference point of world mathematics, they were ready to welcome him with open arms. In 1933 John Von Neumann was among the first to propose the transfer that Gödel definitively accepted, along with his wife, in 1938.

Soon we will meet the two people we saw coming to meet us, chatting quietly. They speak in German and, as they approach, I can distinguish one younger, elegant in his double-breasted suit and the other, older in age, dressed informally with a sweater.

They are Kurt Gödel and Albert Einstein.

Walter Cook, director and founder of the New York University Institute of Fine Arts, loved to talk about the fortune that the United States had been able to host these geniuses, saying that

> *Hitler is my best friend; he shakes the tree and I collect the apples.*

The episode of Gödel became legendary when, in 1948, he applied for American citizenship. Meticulous as usual, he studied the constitution of the United States in depth. The day before the hearing, completely agitated, he contacted a colleague to warn him that the Constitution contained, he said, an error: the Constitution, logically, would not prevent the United States from becoming a dictatorship!

During the interview, after a heated confrontation with the official in charge, Einstein was forced to intervene as a guarantee and only in this way he obtained the coveted citizenship.

In 1940, Gödel and his wife Adele lived a life that was all too quiet: Adele called the Institute a "retirement home" because of the excessive quiet and, for this reason, convinced her husband to move closer to other acquaintances who spoke German.

Gödel, who already did not speak to anyone on his own and very little, even with himself, was a very reserved and shy type. He had very few friends and among these was Albert Einstein.

Einstein, who had by now become an iconic scientist in the scientific imagination and in the society of the time, for his part esteemed Gödel, a young titan of conceptual higher mathematics, so much that he took every opportunity to be with him, taking advantage of the every day home journey to universities to exchange views. Colleagues were not invited to attend these exclusive meetings and had to content themselves with commenting among themselves admitting that Gödel was

> *the only man one who walked and talked on equal terms with Einstein.*

Kurt Gödel, publishing the so-called *Incompleteness Theorems*[14], in 1931, had addressed the superficial plausibility of mathematics by criticizing the formalism of his interpretations on very accurate logical grounds.

He had argued, proving it mathematically and logically, that a proof of axioms held to be true can never be truly proved or held complete within his own system and he proved this by arguing that a in a given system, at least one axiom must be false or unproven .

---

[14] The Stanford Encyclopedia of Philosophy explains the two theorems in this way: "The first incompleteness theorem states that in any coherent formal system F within which a certain amount of arithmetic can be performed, there are statements of the language of F that do not they can be neither proved nor disproved in F. According to the second incompleteness theorem, such a formal system cannot prove that the system itself is coherent (assuming that it is actually coherent)"

This meant that *no* procedure that refers to its own methodological principles could determine truth or falsity.

This conclusion could also imply that a system cannot provide valid results if it is self-referencing. For the universality of the result obtained to be recognized, there must necessarily be a comparison with another system that authenticates the results.

In an even simpler way, in a self-referential system it is not said that the results obtained, based on true premises, are always true, not even in principle.

Gödel's theorem offered no escape: there are mathematical conclusions that are indeterminable even only for principle.

These statements, despite the trauma they caused in the academics of the time, as they questioned the logical basis of the discipline, allowed other developments for physics and mathematics, now inextricably intertwined disciplines: all the advances in mathematics were used for physics; and on the other hand it was often physics that asked help from mathematics to confirm its models.

Till Gödel theories, according to the standard used until then, a "proof" was a series of axiomatic[15] statements.

From Gödel onwards we must consider, in summary, which the concept of provability no longer coincided with that of truth: uncertainty and doubt undermined the foundations of the certainties that had characterized mathematics up to that moment.

This impracticability of setting certain boundaries for a value between truth and falsehood brings us back to the logic that seems

---

[15] axiom: certain principle for immediate evidence and indemonstrable.

to govern *"Quantum Physics"* in such incomprehensible way, uniting the two disciplines in the uncertainty of results and understanding of mathematical and physical theories.

In conclusion, we can formulate perhaps true or perhaps false demonstrations and, study a method to systematically verify their truth or falsity; but this method could never have an end, because the result could never be determined.

Gödel, in the formulation of his theorem, was inspired by paradoxes on self-referentiality that I propose again in the hope that someone can suggest a certain solution.

Let's consider the sentence

The present proposition is a lie

If it is true, then it is false; and if it is false, then it is true.

These are examples of self-referential paradoxes that have always generated confusion and never have a certain solution.

Another paradox, very popular in the Middle Ages, was conceived as an exchange of ideas between Socrates and Plato:

Socrates: What Plato is about to say is false.

Plato: What Socrates just said is true.

Bertrand Russell, mathematician and philosopher, demonstrated that paradoxes constructed in this way invalidate logic and defeat any attempt to build rigorous mathematics on a logical foundation.

Gödel lived as a genius and his skills led him to discover the boundaries of knowledge and truth. Only a philosopher and intellectual like him could have been a mathematician of that value, but he developed a madness that led to an early death.

And above all, as already mentioned, he was among the few to discuss, as a friend, on an equal footing with Einstein.

1934 Panisperna Street Boys

Enrico Fermi (Rome, 29th September 1901-Chicago, 28th November 1954) was fascinated by physics after buying in a market a 15th century book describing nature observation with the ideas of that period. At his entrance exams at the Normal School of Physics in Pisa he showed so much charisma and physical knowledge that examiners asked him to explain his theories. When was a University student, when he attended classes, professors gave him the place to hear his explanations. He completed his course of studies without great difficulty often integrating the subjects to be studied with texts in foreign languages, since there were no publications in Italian on the new emerging *theories of relativity* and *"Quantum Theory"*.

Orso Mario Corbino, director of the Institute of Physics of Rome, in 1923, managed to get him a scholarship that Fermi used to self-finance a six-month stay, in Göttingen, at the school of Max Born, which at that time represented the diamond tip of the most innovative ideas on *"Quantum Physics"*. The studies and theoretical abstractions of Born, Heisenberg, Jordan and Pauli, many times without any real physical meaning, did not convince young Fermi, probably still immature for physics at those levels and his doubts led him to independent studies which led him to work on his own problems of analytical and statistical mechanics.

A year later, Fermi returned to Göttingen, this time thanks to a scholarship obtained from the Rockefeller Foundation, and obtained more satisfactory results both for the intellectual stimuli and for the greater knowledge he had gained in the meantime.

In 1924, from September to December, Fermi moved to the institute directed by Paul Ehrenfest in Leida, where he found a scientific atmosphere that was more congenial to him and began to publish important works of *"Quantum Theory"* that led him to develop the formulation of antisymmetrical statistics at the beginning of 1926, known by the name of *Fermi-Dirac Statistics*, which allow to explain the behavior of some particles called *Fermioni*, from that moment on, and which, together with the *Bosons*, are considered the two fundamental families in which the particles are divided.

At twenty-five he obtained his degree in physics and was almost immediately appointed university professor in the course of Atomic Physics: the address of studies that at the time did not exist at all and was exclusively structured for him by Orso Mario Corbino in Rome.

The success achieved and the innovative ideas on new possible developments led Fermi to move to the institute in Panisperna Street.

In the autumn of 1926, after graduation, he came back in contact with his old friend Rasetti and began a collaboration that in the early months of 1927, was enhanced by Corbino with the assignment of a group of collaborators of remarkable theoretical and experimental abilities, including Edoardo Amaldi, who was the first, followed by Bruno Pontecorvo, Oscar D'Agostino, the great Ettore Majorana and, finally, Emilio Segre.

This led to the formation of a team of organic scholars, whose skills complemented each other.

The institute transformed into a cohort of theoretical and experimental capabilities. An unbridled enthusiasm drove scientists to work more and more, drawing from everyone the best that they had to offer in a healthy competition on any subject of a scientific nature that was raised.

In September, Como held an important international Physics Conference for a Voltian memorial. The prestigious conference also had political promotional aims to officially show the role of Italy among the world powers, thanks also to the scientific development and spiritual renewal of Fascist Italy: sixty-one scientists represented fourteen nations participated. Among them twelve Nobel prizes and the academic standing of the participants made the conference so important that it also overshadowed the recent Solvay Conference.

Some of the participants were F. Ehrenhaft (Austria); N Bohr (Denmark) (Nobel 1922); from Germany: M. Born (who will win the Nobel Prize in the 1954), W. Pauli (who will win the Nobel Prize in the 1945), M. Planck (Nobel 1918), A. Sommerfeld, W. Heisenberg (who will win the Nobel Prize in the 1932); from England: F.W. Aston (Nobel 1922), W.L. Bragg (Nobel 1915), A.S. Eddington, O.W. Richardson (Nobel in 1928), E. Rutherford (Nobel 1909); from France M. the Broglie; from Holland H.A. Lorentz (Nobel 1902), P. Seaman (Nobel 1902); from the USA: A.H. Compton (Nobel 1927); R.A. Millikan (Nobel 1923), all names of great prestige that can easily be traced in any text of Physics.

Sommerfeld and other theoretical physicists produced a demonstration of the new *Quantum Statistics* that explained how to solve problems that until then were insoluble. He spoke with authority and his remarks earned the international respect of the speakers.

He and his research group, then called the Panisperna Street Boys, thanks to the synergies that established themselves between them made a technological leap when they abandoned the spectrographic study of atoms and molecules to begin to deepen the new aspects of the properties of the atomic nucleus which in

1929 Corbino defined as the new frontier of physics, in what became a famous speech.

The evolution of the group reflected the growing scientific stature of Fermi up to its most important contribution to Physics: the fundamental formulation of the *theory of beta decay*, in which he resumed the *neutrino* hypothesis, formulated years before by Pauli, to maintain the validity of the principle of energy conservation in *beta decay*.

Fermi also proposed the idea that protons and neutrons were two different states of the same *fundamental object*, expressing the concept, absolutely new, in which the electron was not pre-existing in the expelled nucleus but, together with the neutrino, was released in the decay process, exactly as it happens for an amount of light when it is emitted due to a quantum leap.

The implications of this mechanism realized an old dream that had so fascinated the ancient alchemists: the transmutation of the elements. In his contribution, Majorana, theorising the existence of the neutron long before the actual discovery of Chadwick, realized the existence of a strong, unknown force that held the nuclear particles together. For his excessive modesty and reluctance, or perhaps for his lack of confidence in the results pursued, as had already happened previously for other famous people. Majorana did not want to publish his brilliant ideas[16] despite Fermi's many encouragement.

---

[16] Majorana was able to produce a relativistic wave equation, following the one proposed by Dirac, seeing what will be called antimatter, for non-charged masses such as the neutrino. According to him, the antineutrino and the neutrino were the same particle that by reversing the orientation of their spin state transformed into one or the other.

Panisperna Street Boys had greatly anticipated the times using conceptual instruments, such as the *anti-neutrino*, which would only be discovered twenty years away from their theories.

Fermi's discoveries were so innovative that the publications on his conclusions, sent to Nature, were rejected because they were too abstract and far from physical reality and so were published elsewhere.

The work on neutron physics continued successfully until the end of 1935 when the group dissolved and the components, in pursuit of different political sentiments, took separate paths.

In that period, Fermi, strong of the experimental experiences acquired until then, expressed the theory that it is the basis of the slowdown of neutrons, useful for mathematical theory at the base of nuclear reactors.

Meanwhile Mussolini acquires more and more consensus and the political situation in Italy deteriorated more and more. Abroad, foreign laboratories were able to obtain more and more funding and material resources, especially for the design of new weapons that could set in motion an industry capable of rapidly supplying more powerful atomic-based weapons. However, despite the incessant demands, it was unable to achieve the minimum guaranteed that a properly equipped national laboratory would not grow.

Fermi refused for a long time the numerous offers that came to him to move to U.S. universities, until 1938 when, following the promulgation of fascist racial laws in Italy that directly threatened his family being the wife Laura Capon Jewish, seized the occasion of the Nobel Prize withdrawal in Stockholm, which he had been awarded for his contributions to neutron physics, to leave Italy.

Already at the time of the withdrawal of the Prize, and then during the celebration ceremonies, he never made the fascist salute, raising a lot of fuss on the media of the time. Then, directly from Stockholm, without returning home, he boarded with his family for United States and Columbia University of New York welcomed him with open arms.

Shortly after his arrival in the United States, Hahn and Strassmann, in Germany, succeeded in the undertaking to achieve uranium fission. While Fermi focused on studying the neutrons emitted in this process, he soon realised that there was a possibility of obtaining a chain reaction capable of producing an enormous amount of energy.

In 1942, December, convinced that German scientists had to be anticipated; Fermi confirmed his adherence to the Manhattan Project for the construction of a nuclear device. The first fission nuclear reactor was built in Chicago shortly thereafter.

Segre also joined Fermi to contribute to the development of the atomic bomb and their contribution was decisive: in fact, the Americans paid tribute to him by heading FermiLab, the most powerful particle accelerator of the United States, a direct competitor of the LHC in Geneva.

Franco Rasetti, on the other hand, had emigrated to Canada and worked at Laval University of Quebec City.

The English had contacted him in great secrecy to obtain his collaboration in the development of their war plans, but Rasetti refused on moral ethics, because he was so convinced that physicists should stay away from politics that they should seem psychologically isolated from reality.

Because of his detachment from political affairs, the scientist ignored that his colleague and friend from Panisperna Street, Bruno Pontecorvo, had meanwhile moved to France to work with Irene Curie and her husband, Joliot-Curie, who were deeply attached to communist ideologies. He probably also let himself be persuaded to embrace these ideologies. Pontecorvo, with Rasetti's mediation, also escaped to Laval University after the Germans had invaded France and arrived there, began to collaborate in the development of secret weapons in Chalk River in Ontario and then in Harwell, but never gained the full political trust of those responsible for research.

At the end of the war, doubts about him were again fuelled when, after a holiday in the Mediterranean in the 1950, he moved with his family to Scandinavia, and then moved directly to Russia, where he spent the rest of his life.

Years later, a senior KGB deserter, also questioned about the role of Pontecorvo, claimed that he had passed information to the Soviets while working on Western nuclear projects.

The question of whether Pontecorvo was a spy was never solved: he certainly played key roles and, for this reason, many spied on him. One of the many explanations was that he leaked information, not spontaneously, but because he was being blackmailed by the Russians.

For his role Pontecorvo was certainly a source of great concern for the blockade of the Western countries.

A separate chapter, among the children of Panisperna Street, must be dedicated to Majorana, on which many stories have been built, perhaps even more than on Pontecorvo.

Even today little is known about his fate and there are no hypotheses about what could have happened to him.

Majorana was recognized as the most brilliant theorist of the group. The only one who alone solved problems with great ease that for others would have been of great commitment and time.

Few months before Fermi received the Nobel Prize, Majorana mysteriously disappeared and nothing was known about him.

Segré and Amaldi suggested that their former colleague had committed suicide. This is a bit hard to believe, as on the last day of his sighting in Italy, it turned out that he withdrew all of his funds from the bank, which amounted to around $ 70,000 at the time: certainly a behavior that had little to do with those who intend to commit suicide.

From some photographic evidence it seems to have been traced to Venezuela in 1950. It is conceivable that Majorana abandoned physics frightened by the possible terrible applications of nuclear science.

Amaldi was the only "boy from Panisperna Street" to remain in Italy. He continued to study cosmic rays and particle physics, becoming a point of reference in the Italian postwar period.

He contributed to the setting up of the National Institute of Nuclear Physics (INFN) and to the creation of CERN in Geneva. The Frascati synchrotron and ESA also see him among the noble fathers.

Fermi's work has contributed in a decisive way to the progress of physics in many fields of daily use and has made great strides in equipment that is now in common use. However, there was a big counterweight: his theories and experiments were crucial to the creation of the atomic bomb, one of the deadliest weapons of human civilization. It is no coincidence that he used the term "civilization": Fermi himself unnecessarily warned those who had to take political responsibility of the great danger that nuclear energy represented, to the point of doubting the ability of a civil society like ours to know how to do careful choices on technologies deriving from the atom.

Fermi died of stomach cancer at the age of 53.

In 1956, in recognition of Enrico Fermi's work, the United States Atomic Energy Commission established the Fermi Award, its highest honor. Among the winners are Otto Hahn, Robert Oppenheimer, Edward Teller and Hans Bethe.

1935 Schrödinger's Paradox

*There is a natural order in
this world and those who try to
overturn it do not end well!*

Haskell Moore (Hugo Weaving) in Cloud Atlas

The New Proposal

Erwin Schrödinger was one of the founding fathers of "Quantum Physics".

His equation served to understand the movement of subatomic elements in the form of a series of probabilities that was dissonant with the ways of seeing classical physics, accustomed to interpret and explain physical reality on certain theories supported by observations and experimental evidence. Most of his contemporary scientists also showed scepticism by considering the new ideas an open violation of the way they think about their contemporary, although there was now a great migration towards the new theories.

The results proposed by *"Quantum Physics"* still today cannot be tested with absolute certainty, because of the probable nature of deterministic verifiability: this approach has always led to great contradictions in the possible interpretation of the results, often generating heated diatribes among the workers.

On the other hand, Einstein's revolutionary theories, which were of a deterministic nature, were also very difficult to accept, because they were also difficult to verify and to understand for the physicists of the time.

Thus the two strands in which physics seemed to have split were now moving in different directions, gaining supporters according to the inclinations of the different scientists; it seemed there was no longer any possibility of convergence.

Scientists were faced with a choice of field: this was the new physics and both roads had in common an intrinsic complexity due both to mathematical formalism that obligatorily had to be used, both because of the conceptual difficulty of having to imagine a new world built on something that escaped everyday logic and that had to be completely imagined from scratch. Being both united by these difficulties of understanding, it resulted in a proliferation not of alternative ideas, but of disputes about the validity of the assertions. Often did not enter into the substance of the subject matter dealt with, or promote an opposition of belonging following the instinctive sympathy towards the one who proposed the original theory.

Above all, physicists could no longer be storytellers of events but were forced to understand the things they were talking about in order to subsequently spread them in the environments they attended.

Often, instead of challenging the validity of a theory on the exquisitely physical-mathematical plane, one preferred to attack the physical constructs trying to identify the flaws, systematically using logic, certainly less expensive than pure physical knowledge.

This method has led to numerous logical paradoxes, often self-referential, but other times of great scientific value, such as in the case of *Einstein-Podolsky-Rosen*, or *EPR* and *Schrödinger's Cat Paradox*. The EPR, or Einstein's Paradox Podolsky-Rosen, is chronologically earlier than Schrödinger's Cat, but we will deal with it later, because it proposes logical analogies with other subsequent experiments.

We will therefore only talk about *Cats*, who have entered into the common science imagination.

The Observer in Quantum Reality

In the process of analysing quantum phenomena, as we have already noted and, especially in the observation of phenomena related to the effects produced by *"Quantum Mechanics"*, a highly critical and debated aspect was the role played by the observation process, meaning any element, human and/or instrumental, capable of determining or observing experimental results or a quantum physical system.

In the every day world, because of the large masses at stake, this problem does not have a significant impact.

The *physical system* idea has different meaning: any set of elements, from a single atom to the entire Universe, can be considered a physical system. Every element that makes up a system can also be another physical system in its own right: for example, an electromagnetic field or a quantum atom, although they are part of larger systems that obey well-defined laws, they are also systems that obey laws often in opposition, or however different, from the container system.

A system can be completely autonomous from the world around it both physically and logically, but it can also not be, thus involuntarily involving the observer.

By observing a system we acquire knowledge of its mechanism of operation, or revelation, which is pre-existing to our observation.

Until the laws of *"Quantum Mechanics"* were determined, this entire logical construct did not have to exist, or in any case did not create obvious problems. The results of the experiments were independent of the presence or otherwise of an observer and the observer did not influence the system with his presence.

The physical elements and systems were not substantially modifiable and maintained both their characteristics and the manner in which they had been achieved.

The study of "Quantum Mechanics" has completely changed this philosophy: from the very first moment in the study of the subatomic world, the observation and the way in which these measurements were made became decisive, as the effects of an external observer were well visible in the results of the measurement.

Did these premises call into question the repeatability of the same experiment?

In different places would it have produced different results?

And if carried out in the same place by different people, would it have produced different results?

Or, even more intriguing, there would be another question: would an experiment observed by a being other than man, with different sensibilities and peculiarities produce the same result?

Or the continuity of the result is confirmed only by the presence of the constant human observer?

These questions would pose a major question, in particular, to the Bell experiment, which we will see later.

Above all, the revelation phase has proved to be the most critical because, in certain well-defined cases, it could, consciously or not, lead to the creation or destruction of parts of the system itself, which, from that moment on, would become another system with its own reality completely different from that of the previous one, which should be studied not only as a system evolved from a previous one but, above all, as a completely new and existing system only from the moment the observation starts.

The inherent problem is that this new system was created by our observation.

This reasoning, for example, could lead to another paradox or another logical experiment: if we apply this reasoning to our Universe, then it would be legitimate to ask, for extremely speculative hypotheses, whether it has existed before, whether it has always existed, or whether, paradoxically, may be the result of any observation?

It is a hypothesis of a mental experiment that cannot be discarded because it does not admit a real contradiction: neither the Big Bang itself, the most currently accredited theory of the birth of the Universe, nor the theory of the Big Bounce, another emerging theory that finds many supporters, can offer more comprehensive scientific explanations.

In this totally unknown of the initial zero moment, there is no reason why even a *creationist theory*, that has always considered as the only difference, however not demonstrable, the voluntary act of the birth of the universe.

So the observation process is a cornerstone of the experiment and unexpectedly opens up an ontological question[17], since it involves in the phase of physical research an acceptance of the balance between our *"humanity"* and the ability to become

---

[17] Ontology is any reference to the knowledge of philosophical "being" that is assumed.

creators of the experiment and its result ourselves, because observers.

This was the consequence of the conclusion that scientists had come to, as soon as they began to do the first experiments in *"Quantum Physics"*.

The presence of the observer and the ways in which he acquired the experimental data made it impossible to maintain the initial conditions of the system, whatever the observation system was.

This is still valid today in the observation of the subatomic world, because the masses of any observation system irremediably influence the result: even just looking in the direction of the event influences the system enough to modify the result.

Due to these reflections, *"Quantum Physics"* also turned towards a metaphysical understanding of phenomenology, just as it was at the beginning with Leucippus and Democritus, who were our starting point; moreover, it redefines, re-proposing it on a new scale actually experimented and also mathematically verified, the old hypotheses of realism that had belonged to scientists until the advent of *"Quantum Physics"*: there really is a reality that is completely outside the entities they observe, and it is a reality perfectly describable, just as scientists did up to 1800 through the enunciation of verifiable theories.

If this were true, we could go so far as to say that a system has its own awareness that manifests itself when the observer's intellectual awareness exists.

In other words, more simply, a system provides certain results because we are the ones who are observing it.

The certainties we were used to and the very concept of realism crumbles because a unique and objective representation of reality disappears and the observer himself becomes part of the experiment.

In this way the system from which we started has regenerated itself in a new system that involved the same observer, unaware and, establishing new rules of interchange. These reflections have significantly changed the study of quantum phenomena.

In the logic of *Schrödinger's Cat Paradox*, the observer plays a decisive role.

1935 Schrödinger's Cat

Following the Solvay congress of 1927 and the great multiannual discussion between Einstein and Bohr, Erwin Schrödinger also sought a way to understand the new experimental results that *"Quantum Physics"* experiments now produced continuously. Schrödinger, despite having proposed his equations for this purpose, believed that beyond the appreciable results he was able to achieve through them, there was still room for deterministic interpretation.

The idea, therefore, was a logical experiment, not feasible in everyday life, which had to support Einstein's thesis, convinced determinist: this experiment, which took the name of *Schrodinger's Cat Paradox*, instead of revealing the problems with *"Quantum Physics"*, contrary to his intentions, it became one of the paradoxes that promoted it better.

Because Schrödinger used a cat to declare the experiment, we will never know. He probably had an open argument with these animals that he probably risked, every time we talked about this paradox, not getting away with it.

Let us repeat with a lot of misunderstanding, that the cat paradox is an unreliable mental experiment in reality: therefore, no living being dies.

Conceptually, it is easy to understand and logical in its adherence to physics that initially intended to refute and absolutely not predict any result first to begin.

In this type of experiment, or better this reasoning, it is essential that the experimental method and the results obtained are extremely logical.

Schrödinger's paradox establishes a connection between the world of elementary particles, protons, neutrons, electrons, photons, etc., which obeys the laws of probability with all its contradictions and the macroscopic world governed by deterministic laws.

The assembly of the experiment would be quite simple: a black box closed by a lid that does not allow you to see what is going on inside; inside the box a vial containing a poisonous gas with an electrical opening mechanism activated by an external pulse; a single uranium atom with a half-life of one hour, a Geiger counter. Connect the counter to the opening device of the vial so that, if the presence of radioactivity is detected, the switch that opens the vial activates. Of course, we put a cat in there, any color, and we close the box.

Let the appropriate time pass, let us say at least one hour, to give the time, possibly, for the uranium atom to decay, without any influence from the observer.

According to the logic of the wave function, after one hour the radioactive atom will be in a state of decay or non-decay. If it decayed, the opening mechanism of the vial activated and the gas contained poisoned the cat by killing it. If it has not decayed, the cat is alive and the gas is still in the vial because it is still closed.

For the observer, until the box is opened, the cat could be alive and dead at the same time, because he is unable to determine its state. When the observer decides to open the box to examine what happened, he intervenes for the "measure" and determines the equivalent of the collapse of the wave function in one of the two states.

Until then, the two quantum states have always been superimposed.

This is a key aspect of the Copenhagen interpretation of *"Quantum Physics"*: it is not just that the scientist does not know what state he is in, but rather that physical reality is not determined until the act of measurement takes place. Somehow unknown, the very act of observation is what determines the situation in one state or another.

Until that observation takes place, physical reality offers all possibilities.

According to everyday logic it is not possible in the real world to be alive and dead at the same time!!! Instead, according to the quantum interpretation of Copenhagen, this is possible: the cat is simultaneously in two different states.

This paradox has become the cross and delight of many physicists over time, so much so that many have tried to refute it, many have accepted it for what it is and many have commented on it according to their own humour.

In fact, when Stephen Hawking spoke of it to describe the degree of unbearability that this transmitted to him, he liked to say that:

> *When I hear of Schrödinger's*
> *cat, I reach for my gun.*

This expression also coincided with a widespread view among many physicists, due to the many contradictions that the experiment produces.

The most relevant problem of the paradox is that it must be related to the dimensions of *"Quantum Physics"*, which operates only on a microscopic scale of atoms and subatomic particles and, not on a macroscopic scale of cats and vials of poison. We use dimensional elements of our world in the thought experiment only to make it easier to understand what is happening.

Furthermore, as long as the box remains closed it is impossible to know if the cat is alive or dead, so for normal, non-physical people, it is in an indeterminate state, in which it is both alive and dead and, not in a state superimposed. They are technical adjustments for use and consumption of *"Quantum Physics"*.

The experimenter, who acts as an observer, as we can well understand, has a probabilistic role because he can at most hypothesize that there is an equal 50% probability of finding the cat alive or dead.

The event followed its natural unfolding, but what happened has not yet been determined.

The observer, to know the fate of the cat, must open the box and look inside: in carrying out this act of opening, the certainty of observation is acquired and the fate of the cat is determined, which until that moment is simultaneously alive and dead, with same probability.

If in macroscopic world all this makes no sense, when we talk about the particles of subatomic world, the extravagant idea of thinking that these, at the same instant, can assume different states turns out to be successful in justifying the illogical behaviors that distinguish them. It is therefore possible to say that the world of atoms is really a world apart, where things happen in a way that our brain refuses to accept.

We can never be innocent again

> – Dr. Einstein, because when the mind of man discovered the structure of the atom, you scientists were unable to find the political means to prevent the atom from destroying us?
>
> – It's simple, my friend. Because politics is more difficult than physics.

Moe Berg: the Probable Heisenberg Killer

Shortly before the start of the Second World War, on August 2, 1938, Einstein and other scientists wrote a letter to the American President Roosvelt in which he pointed out the risks that were running since the Germans had gained an advantage in the nuclear arms race, especially because with the invasion of the Czech Republic they had secured the only uranium mine in Europe.

Unlike the military leaders, the dangers seemed much more real to scientists who signed the letter to the politicians, because they realized that in addition to the main material, uranium, the Germans Hahn and Strassmann, in 1938, carrying on the 1934 Fermi's studies, were able to achieve nuclear fission for the first time. Above all they could count on the Nobel laureate Werner Heisenberg, the best theorist in the world of atomic physics.

The letter, which in the following times would have led to the two bombs dropped in Japan, at first, produced such an agitation as to lead the Allies to try by all means to prevent the Germans from achieving the result before them.

To do this they relied on a spy game worthy of the best novels of the genre. Only that this time it was all reality.

A few weeks before the letter was compiled, Einstein himself had met Heisenberg on the campus of the University of Michigan, where he had come to lecture, invited as a distinguished guest. All the best minds who were currently studying atomic physics were present to share ideas and projects. The meetings also served as moments to make or strengthen friendships. Heisenberg was Samuel Goudsmit's guest, an America physicist that was his old friend, with whom he often had correspondence. Recently Heisenberg had become famous for his *Uncertainty Principle*: it was proposed by Einstein as candidate for the Nobel Prize.

At the meetings there was also present Enrico Fermi, a young promise of Italian physics, who in a few years would have carried out the first nuclear chain reaction that served to enrich the plutonium necessary for the construction of atomic bombs.

Goudsmit, in addition to being a valid scientist, was probably already working on behalf of his own government, in order to understand the state of research of scientists from other countries.

One of those present, subsequently speaking with a reporter reported that

> *That Goudsmit guy, he talks about physics, but he also talks about a lot of other things.*

In addition to science, the topics that were discussed were interested in the events of that period and many of the witnesses reported how at one of the receptions, Heisenberg had answered, to those who asked him how he could remain in a country in which it was gradually establishing the dictatorship, of wanting to contribute so that Germany was led back to a more rational choice. Oddly, he added that he was forced to return, letting us suppose someone had invited him to remain in United States.

Two of the boys from Panisperna Street, Fermi and Amaldi, also attended the receptions.

Amaldi, while observing his friend and colleague Enrico Fermi and a little distance, secluded and solitary Heisenberg, commented in a low voice with a colleague:

> *See Fermi, see Heisenberg, sitting in a corner. Everyone in this room expects a big war and the two of them will lead fission work on each side, but nobody says.*

Unfortunately Amaldi's prediction came soon after.

In the last months of 1942 Goudsmit was warned by Hans Bethe, an immigrant physicist, that Heisenberg would go to Switzerland for conferences. He suggested, unsuccessfully, to kidnap him, but the idea had made its way to General Leslie Groves, military director of the Manhattan Project. By coincidence, since that time the Germans no longer published scientific articles on the atom and for the Americans this sounded like "no news, bad news". They feared that the silent silence would hide the results they were getting.

And so, desperate times call for desperate measures, General Groves made contact with William "Wild Bill" Donovan, head of OSS, the Office of Strategic Services, the precursor of the current CIA, so that Heisenberg could be stopped in any way, including by taking him out into the field.

Wild Bill commissioned Colonel Carl Eisler to complete the mission by giving him a complete blank slate on the modalities, over a virtually unlimited budget.

The plan was simply adventurous: a dozen soldiers were trained at a secret OSS base in Maryland. The idea was to transport them by air to Switzerland, from where they would then enter Germany to kidnap Heisenberg and sneak him back to Switzerland, from where he would be taken by plane that would fly over the Mediterranean.

At a predetermined X point, a submarine would be found waiting, which would then collect the parachuted scientist from the plane. All against his will, of course.

Just before the start of the operation, the head of the OSS in Switzerland strongly opposed and stopped the operation because he believed that the Swiss would join the Germans if they had learned that someone had violated their national space, which was neutral.

The fear that the Germans could achieve the goal first continued to worry American military summits, even more when Nazi propaganda minister Joseph Goebbels began bragging about the media having a "miracle weapon" in the yard at an advanced stage of development, beyond a fantastic "uranium torpedo".

Groves then tried another card and literally brought in a former professional baseball player, Moe Berg, who was very well prepared: in fact, he spoke many languages, including German and Japanese, he had graduated from Princeton in Letters and in Jurisdiction at Columbia.

In 1942, after abandoning his sports career, he decided to serve his country by joining OSS and becoming one of Donovan's leading men.

He was sent to Italy to try to find out what was known about the German project in scientific circles. Meanwhile, Groves appointed Goudsmit scientific head of Alsos mission, of which we have already mentioned.

In 1944, Goudsmit told Berg, on Donovan's orders, to go to the conferences Heisenberg held in Zurich, granting him permission to kill him if he felt that the German scientist had already arrived at the weapons expected by the German military.

It was a suicide mission because SS agents closely monitored every movement of the scientist.

Berg, who spoke German, studied in depth the texts that he was able to obtain on atomic energy and went to the auditorium carrying a Beretta pistol, posing as a physics student.

He sat in the front row and furiously began to take notes. At the end of the conference, mixed in the crowd that had approached the board to read the equations, heard Heisenberg to say:

> *We are losing the war, but*
> *how nice it would have been if*
> *Germany had won.*

Berg approached him again by attending the next dinner and, accompanying the physicist to his hotel, as he walked beside him, gave up the intention to kill him, because he realized that it would be a useless action since, in his opinion, the physicist did not represent a real danger.

But one question remains about that walk they took together: however skillful Moe Berg might have been, how could he have fooled a physicist like Heisenberg with his knowledge of virtually nothing?

What they really said, only the two men knew, the fact is that Heisenberg remained alive.

And let's not forget: Berg was Jewish, so he had every reason to kill Heisenberg just because he was a Nazi!

The Alsos mission, shortly after, when it came to capturing the German scientists, confirmed the validity of its choices, because they found only one project in embryo and largely late.

When Heisenberg was arrested, he had his picture framed on his desk with Goudsmit, taken years earlier at that meeting in Michigan.

After the end of the war Goudsmit and Heisenberg tried to reconnect unnecessarily and, in the 1976, Goudsmit, in his praise of Heisenberg, commented saying "I had lost a friend".

Today, reviewing those episodes, many think it was a stroke of luck that the Allies didn't kill Heisenberg. His role remains shrouded in the fog of doubt: has he really failed to bring Germany to the atom or has he deliberately slowed down the programme?

Probably someone else in his place could have caused unimaginable damage.

When the Alsos soldiers reached Bavaria, they collected samples of whatever happened at gunpoint to send them home to scientists who would look for traces of radioactivity.

Together with water samples, they also sent, as a joke, bottles of French wine that they had confiscated, warning scientists to look for radioactivity in those too.

Scientists without any sense of humour examined the wine and, in fact, they found it radioactive.

By its very nature, the wine contained a low level of radioactivity due to the soil on which the vineyards were grown. Scientists, who did not know it, forced the men in the territory to collect samples of other soil to verify the cause of the radioactivity.

Moe Berg remains today a legend whose deeds have not yet been known all.

August 6, 1945: Hiroshima

On the 16th of July 1945 in Alamogordo, New Mexico, the *Manhattan Project*, to which the best scientists emigrated to the United States from all over the world, led by Robert Oppenheimer had contributed, to reach the goal of realizing and detonating in the *Trinity* test the first plutonium bomb, code name *The Gadget*.

On the morning of August 1945 to 8.17 *"Quantum Physics"* reached the point of no return and, as for the life of a human being, it passes from the carefree phase of youth, lived until then romantically and adventurously, to a dark awareness of its strength and its new reality. Its wonderful history changes immeasurably: it is the birth of a new world that has in it the awareness of the possible end of human civilization and of the survival of our species.

The USA's strategic Boeing B-29 Superfortress, called Enola Gay by the name of the pilot's mother, and dropped the bomb *Little Boy* on the Japanese city of Hiroshima. Three days later, August 9th, Nagasaki was also hit by *Fat Man*.

The number of victims, as a direct consequence of the explosion, was more than 200.000, almost all civilians.

The bomb dropped by Enola Gay, was three meters long and four tons of weight, and exploded about 580 meters high, developing a power of 12.500 tons of TNT.

Hiroshima was completely destroyed by the reaction of the split of uranium and plutonium nuclei that caused a temperature rise such that the whole area melted for miles from the point of the explosion.

The next shock wave travelled at a speed of almost 3.000 meters to the second and shaved to the ground everything that had survived the heat within a radius of 800 hundred meters from the point of the blast, expressing an inertial force of seven tons per square metre and, causing the immediate death of nearly 70.000 people, to which were added 100.000 who died in the following months due to the injuries reported and to the radiation that developed.

Both with Enola Gay, who had dropped the bomb, there were two other B-29s, *Great Artists* with scientists on board to observe and analyse the data and *Dimples 91* equipped to document the military operation with footage. Everyone in the stories of the following days told the mixture of horror and charm of what they had experienced in those few moments seeing the destructive effects and the gigantic atomic fungus that had formed.

The Enola Gay pointer, Tom Ferebee, let himself go to a *"My God, what have we done"*, while the commander and pilot, Colonel Paul Tibbets, calmly commented on the onboard radio *"Gentlemen, you have just dropped the first atomic bomb in history"*.

One of the effects that occurred in the following days was the almost total attenuation of the intensity of light that lasted several days and would be used in subsequent science fiction stories to develop the idea of nuclear winter that could occur following a hypothetical third world war.

It seems that many protagonists of these events, civilians, scientists and military, have developed a syndrome of guilt and remorse for what they had done, which lasted and conditioned their remaining lives.

Atomic Bomb Explosion: the day after

Following the launch of the bomb, Harold Jacobsen, one of *Manhattan Project*'s scientists argued, at first, that there would be no life forms in the affected places for at least seventy-five years.

Instead, both in Hiroshima and Nagasaki, trees about two kilometers from the epicenter of the explosion began to sprout already in the following springs and in subsequent studies somebody also talk of surviving trees within a radius of 500 meters from the epicenter. This exceptional event is in contrast with the idea that nothing could have survived.

Stefano Mancuso, plant physiologist and director of the International Laboratory of Plant Neurobiology at the University of Florence, wished to express his opinion in this way:

> *In retrospect we know that this may be due to the fact that some buried parts of the trees have been protected by the layer of earth, or because on the non-radiated side, protected by the thickness of the trunk, something has survived. Those specimens are reborn because plants are not a "single organism", like animals: they have instead evolved in a scheme that we could define as "modular" to survive the predation of animals capable of feeding up to 90% of a plant. With a simplification, we could compare them to insect colonies.*

These resilient trees were later identified with an inscription, surveyed and registered. They have been called *hibaku jumoku*, a surviving tree. In the only Hiroshima there are about one hundred and seventy plants of thirty-two different species. The tree closest to the epicenter is a weeping willow. Hiroshima residents share the seeds of *hibaku jumoko* with the world in a symbolism of destruction and rebirth.

J. Robert Oppenheimer, head of Manhattan Project, although he had been sympathetic to Fermi in advising to strike Japan, during the Gadget atomic test in New Mexico, defined himself: *I became Death, the destroyer of worlds* and, immediately after the outbreak of the Hiroshima bomb, in the throes of despair that also affected many other fellow scientists, he commented: *Physicists have known sin.*

Oppenheimer further contributed to the development of *"Quantum Mechanics"* and refused to work on the hydrogen bomb. Instead, he made a significant contribution to understanding the *quantum tunnel effect* and the collapse of large stars due to gravitational force.

This confirms the convergence of studies between cosmology and the atomic world.

Szilárd, another scientist who signed with Einstein the letter to President Roosevelt to urge the development of an atomic weapon, later, realizing the danger and the great destructive force of the bomb, tried to convince the other scientists and the United States government not to use the bomb.

"Quantum Physics" From 1950 onwards

*If I could remember the names
of all these particles, I would have been a botanist.*

Albert Einstein

The World of Physics after the War

It is clear that the great acceleration that began in the early 1900s and lasted throughout the first half of the 20th century, to put it in a description of a motion that physicists will well understand, has gradually flattened into a uniform rectilinear motion. For those who are not physicists, it means that the proposals for new ideas have gradually lost their vigor until they level out into a daily normality.

The beginning of the century had brought the two great scientific revolutions with cognitive changes like never before in human history: the relativist and quantum revolutions.

The dates must remain indelibly etched clear in our memory: *Special and General Theories of Relativity* presented by Einstein in 1905 and 1915 and *"Quantum Mechanics"* thanks to Heisenberg in 1925 and Schrödinger in 1926.

When the war was finally over, the scientists, exempted from having to concentrate on improving the war effort of the military, finally returned to objectives more suited to them: to compete in the field to search for those truths yet to be revealed.

The war had brought to light new protagonists who wanted to direct physics towards new goals.

EPR Paradox, Bohm and the Hidden Variables

Albert Einstein was always a convinced assertor that any physical theory should have fully described the phenomena of the physical reality that surrounds us and, therefore, *"Quantum Mechanics"* should also have belonged to a more complete deterministic theory.

All the scientists who shared his thought were convinced that a more complete theory could be found, starting, for example, from the methods used by Boltzmann and Maxwell to explain the laws of thermodynamics, so absolute pressure and temperature represent the average of the values of the properties of molecules. In this way position, speed, amount of motion, kinetic energy... would have been perfectly understandable properties in a deterministic mechanics, because causal and real. When analysing a thermodynamic system, the individual parameters describing the state of the individual particles were not used, but a method was used that mediated their state and behaviour.

These variables, even if present and real, were not used and were called *Hidden Variables* of this level.

We need to take a step back to 1935 to understand how, many of the current developments, result from a hypothesis that *Einstein, Podolsky and Rosen (EPR)* designed to demonstrate *"Quantum Mechanics"* could not be considered a complete physical theory and that there had to be *hidden variables*, like the ones we saw just above, not yet identified, that would have been able to complete it. The reason why the *Paradox EPR* is proposed chronologically afterwards is because, together with the *Bell*

Theory and the *Aspect Experiment*, it represents a context in itself that will last until the 1980s.

All three are particularly complex both in the preparation of the experimental phase and in the conceptual formulation that you want to achieve.

To facilitate their understanding, I will simplify the descriptions as much as possible, trying to keep both logical and conceptual validity intact.

Albert Einstein, it should be clear by now, was not convinced of the validity of the Copenhagen interpretation. The main cause of this disagreement was the inviolability of two fundamental principles for him: realism, for which elementary physical objects retain well defined properties even when not observed and, the location, so these objects can only be affected if in their immediate vicinity there are objects that can modify their properties, therefore not with "remote action".

This stubborn opposition had deep roots in his theories of relativity, of which the locality is an indispensable hinge and in which time and space are directly connected to the observer.

Consideration not to be underestimated: Einstein's theories received more and more confirmation over time, and were able to anticipate exactly the results expected to be achieved before they were confirmed by the experimental observation.

In relative terms, the location is the clear explanation of why the effects are caused and not vice versa.

On the other hand, Bell and Aspect, one of Bell students, ventured into demonstrating the non-locality, indispensable for

Bohr's probabilistic **"Quantum Physics"**, Heisenberg and Schrödinger's equations.

In short, the crux of the problem was precisely in this contradiction between a theory that predicts and confirms and a theory in which we cannot predict the result and we will never receive a certain confirmation, but, as a result, a probability.

EPR says that a physical theory is complete only if it has considered all the physical elements of reality, they say, that every physical quantity should have a value that can be predicted exactly before any measure is made.

Einstein, Podolsky and Rosen, this time, took inspiration for the idea of the paradox from the *state of entanglement*, i.e. of entanglement, predicted by Dirac in the 1928, so that two microscopic particles that interact for a period of time and with certain modalities in a closed system, even if separate will continue to behave as particles, but unexplainably they will have memory of this interaction instantly conditioning each other, in fact constituting a single system until they come into contact with other particles.

This kind of entanglement, in everyday life, is much more frequent than one might think: if I divide a piece of pie into two parts, one higher and the other lower, and give one of the two parts to a friend, the moment I eat my part I will instantly know which is the second part that will eat my friend, wherever in the universe he is.

Let us now imagine that our system consists of a source that emits two electrons A and B in mutual entanglement, which are separated and directed towards two different measuring instruments. When we measure one of the two, because the second electron is in entanglement, i.e. together they form a single system that binds them indissolubly, according to *"Quantum Mechanics"*, i.e. that they are always correlated everywhere and that the second observed measure is predictable instantaneously, we will also know the measure of the other not observed and vice versa.

This implies that the observer, by performing the measurement, according to what is stated by *"Quantum Mechanics"* would lead to the collapse of the wave of the first and at the same time also of the second, without even the interaction of the observer.

The entire single system, but also consisting of two other systems, now distant and different, undergoes an instantaneous change of the two distinct and separate distant realities between them.

This theory, for Einstein, was impossible because it clearly violated the principle of a *strong locality*, experimentally verified and predicted by relativity, so distant objects cannot affect each other instantaneously: an element is directly affected only if it is in the immediate vicinity, so that a physical process cannot have an immediate effect on physical elements of reality of another event separated from a space, since a cause of interaction can at most propagate at a finite propagation speed, which at most can be that of light, as established of limited relativity.

Consequently, any theory that violated relativity, for Einstein, was inconsistent.

After this experiment, Einstein defined quantum prediction as a *phantom action at a distance* and concluded by pointing out that *"Quantum Mechanics"* lacked something and, therefore, was an incomplete theory.

Subsequently, the paradox was addressed by the same EPRs who used it to hypothesize, although *"Quantum Mechanics"* produced correct results in many experiments, so that the statistical approximation could explain the results.

In other words, the statistical probability, which correctly predicts a probable result, does so because, being incomplete, it can only produce a probable result.

So there must be other *Hidden Variables* that, unknown to the observer, afflict the physical elements of reality, both for the U*ncertainty Principle* and for the Complementarity *Principle*, that by imposing these limits originate the effects that *"Quantum Mechanics"* can only interpret at a probabilistic level.

Regardless of how the paradox was implemented, the conclusion that EPR wanted to reach was that *"Quantum Physics"*, according to the Copenhagen Interpretation, provided valid but probable results because of some *Hidden Variables* that led observers to reach those results in an unconscious way.

There was still a lot of work to do to track down the *Hidden Variables*.

The controversy erupted as gasoline poured on a fire, leading the contenders to divide, once again, into two opposing currents of thought. The first was linked to the school in Copenhagen, supported more and more unconditionally the view that the results were likely and indefinite in relation to reality, and objected that the underlying assumptions of the paradox from the initial stage of the approach were not in line with this interpretation. The second current, which was based on the conclusions reached by EPR, considered *"Quantum Mechanics"* incomplete and considered that additional parameters had to be sought to explain the strange correlations that appeared to emerge between distant systems.

From this second stream of thought came the new theory of the *Hidden Variables*, which had among the fathers the same Einstein, Podolsky and Rosen. The presence of additional parameters, according to the scientists who support this theory, would produce a double advantage: *"Quantum Theory"* would be complete and would solve the paradox; In addition, it would bring reality back to the determinism that is incumbent upon it.

We are at a crucial point and suited to a question that since the post-war period someone had begun to ask themselves subtly: physics had ended in a dead end?

Bohm's Hidden Variables

David Bohm (Wilkes-Barre 1917 - London 1992), was an American physicist and professor first at Princeton University until 1951 and later at Birkbeck College in London. Richard Feynman (New York 1918 - Los Angeles 1988), was an American physicist, very exuberant and outgoing who had participated in the Manhattan Project, in 1952. They met in Belo Horizonte in Brazil.

Bohm was teaching at the University of Sao Paulo after he had been forced to leave Princeton University and the United States due to the McCarthyist paturnias, while the usual bursting Feynmann was in Rio, because he had taken a year off and had decided to spend it relaxing among beautiful girls and bongo lessons. Bohm, as was his custom, was going through a difficult time because he felt in exile far from his country and also distant from the Physics that mattered.

Feynman, however, expressed admiration and interest to him for his innovative and original ideas about the conception of *"Quantum Mechanics"* and managed to awaken him from his scientific torpor.

Perhaps, for the first time, there was a possible real alternative to the Copenhagen Interpretation: even Bohm, using the works of Einstein and de Broglie as a starting point, did not share the probabilistic approach of the quantum world to express the results of the subatomic world. With his theory of *Hidden Variables* he intended to settle accounts with the paradoxes of Schrödinger's cat, with EPR, but above all he intended to provide a deterministic answer of reality to definitively solve the still unsolved problems: the overlap of states in the macroscopic world, explicit reference to *Schrödinger's cat paradox*, *the collapse of the wave function* and others, such as *the role of the observer*.

The first statement considered particles *stable* in predetermined positions and subjected to a *quantum potential*, a force that operates like electric and gravitational forces. This potential connects all the elements of the Universe and is independent of the place where the measurements and experiments are carried out; moreover it is not subjected to the limits of relativity and therefore transmits information instantly even at enormous distances. Bohm also expressed the innovative idea that *"Quantum Physics"* was an apparent and superficial part of a *holistic reality*[18].

It was the birth of a new *Mechanics*, called *Bohmian*, an alternative to *"Quantum Mechanics"*, not Relativistic and, above all, obtained with a simple explanation that solved the well-known and long-standing problem of measurement and observer.

In one only shot there was a theory characterized by a non-locality that reconciled Einstein's locality and the quantum non-locality, cancelling the starting hypotheses of both and bringing them together in a theory that summed up both: it was a theory that was apparent conflicting with both *Relativity* and *"Quantum Mechanics"*, treating them both as incomplete.

It is evident how the poor Bohm had put himself against almost the entire scientific world of the time and, not only for his conclusions so out of the ordinary.

---

[18] Holism from the Greek ὅλος "whole, whole, total - in this case the term is used as if the whole Universe in every single and tiny part contributes to being reality.

David Bohm was a formidable theoretical physicist but he also cultivated other interests in humanities field and this, in a world as conformist as that of physics had become, led his colleagues to consider him a heretic to be looked upon with suspicion.

The true is that Bohm had an open mind, ready to look elsewhere, and his preparation made him intolerant of the extreme and often incomprehensible paradoxes of *"Quantum Mechanics"*. He also took up ideas now considered obsolete, since he believed that the Universe could be perceived as a single continuous flow and that this feature was an intrinsic property of matter.

Therefore, for him matter was not a trivial agglomeration of autonomous particles and therefore he was sure there was a common basis between *"Quantum Theory"* and *Theory of Relativity*, both being theories belonging to the same Universe.

Bohm was not convinced that an observer can influence consciously the experiment, rather he was also willing to imagine that there was something beyond the normal logical-scientific capacity that in some way, similar to the quantum leap phenomenon, beyond logical thinking, would justify why suddenly a new idea may arise without connection to previous experiences.

Quantized thinking and its quantum leaps are the new ideas that seem to arrive suddenly but which would all be competing elements to a single formulation of the logic of a process.

All without the individual being aware of it.

There was enough for none of his colleagues to invite him to breakfast.

Bell's theorem and experiment

In 1964 John Stewart Bell (Belfast 1928 - Geneva 1990), a Northern Irish physicist, stated without possible misunderstanding with his Theorem, which I express in the simplest and easiest form, that there is no theory that satisfies the *Location Principle* and the *Hidden Variables Principle* and can provide the same results as *"Quantum Mechanics"*.

With this statement Bell did not deny the *Hidden Variables*, but he did not give them a determining value and at the same time reaffirmed the *non-locality*.

It was a qualified *endorsement* to *"Quantum Mechanics"* that provided a contribution to his refusal of the locality, even coming to interest philosophers who saw new fields of investigation opening up with the ideological fluctuations of modern physics.

In fact, there were all the characteristics to imagine a different reality that surrounds us, with possible *Hidden Variables*, objects consisting of particles that probably exist and realities that were not visible because they exist but not verified until measured. This was only part of the possible interpretations.

Bell's experiments proved irrefutably the non-locality of *"Quantum Mechanics"*.

In fact, no one is willing to give up their ideas and therefore checks and cross-checks were debated, trying to definitively settle the questions of a Physics that after the controversial Bohr-Einstein, had to decide again which way to follow: if the indeterministic, probabilistic and random, or deterministic, localist and causal.

These two possibilities, if it were not for Bell, seemed to be accompanied by a third way, Bohm's theory, which led to mediate the better of the two previous ideas by introducing a new concept of the total Universe in which everything is mutually linked and influenced.

Physics is no longer the observation and discovery of natural phenomena, but an infinite list of unverifiable laws and elements contested in turn by the contenders of the moment.

Occam's rule too,

> *an injunction not to make more assumptions than you absolutely need,*

is no longer certainly applicable, but it probably becomes applicable to a natural universe that tries to be one from the subatomic world to the astronomical sizes of celestial bodies.

For the first time we have to ask ourselves the question: what if the most obvious explanation is not the true one?

Unfortunately it is once again evident that even scientists, in this case physicists, are men with their drives and weaknesses; they lack a breadth of views that lead them to an objectivity of judgment or perhaps they completely lack a self-irony that grants the opponent the ability to express a different idea and refute it with the smile of an open mind.

Thus Bohm and his ideas will remain at the disposal of those who want to risk intellectual exile and in the universities they will continue to study a probabilistic and romantic physics, only because there is a teaching class that is transmitted from generation to generation the initial Copenhagen mantra.

It is good to remember, however, that people like Penrose shared many aspects that had struck Bohm and it is not to be underestimated that Bohr too, with the choice of the Yin Yang emblem for his honor and the respective phrase *"contrary sunt complementa"* could have sensed something like this a long time ago.

These are the facts, everyone can draw their own conclusions, avoiding attacking those who think differently.

Scientists too are men who let them be fascinated by what they believe in and, unfortunately, they believe in what gave them the security of being scientists.

Or not? Or is this the difference between being a scientist and being a genius?

Alain Aspect Hot Shot

From David Bohm in 1950 and John Bell in 1960, the scientific community understood the real great innovation of EPR, but it was in 1980 thanks to experimental results by Alain Aspect that they set a tombstone on the quantum debate by Einstein and Bohr.

Non-relativist *"Quantum Mechanics"* with Bohm had found *causality and determinism, rational, clear, exact and experimentally equivalent* and there are also physical phenomena, *entanglement*, which escape the locality without relativity being affected.

All this, however, only thanks to the contribution of the *Hidden Variables*.

Bell had contradicted Bohm and all returned to the previous impasse of Bohr and Einstein.

The experiments carried out later, although increasingly sophisticated, never managed to definitively nail down the local realist interpretation except by introducing, however, possible additional artifices.

The conclusion seemed to state that local realism cannot be confirmed in the light of these experiments and therefore we no longer have to continue presenting it.

Until …

1982 Alain Aspect

Alain Aspect (Agen, June 15, 1947), French physicist, was among the first to be convinced of the validity of John Bell's theorem. He has persisted in demonstrating that *"Quantum Mechanics"* has an indispensable need for non-locality, thus destroying Einstein's faith in *locality*.

So Aspect completed this triptych of investigations in 1982 by realizing an experiment that repeated Bell's tests with greater precision and without the many compromises of the original experiment.

With this experimental confirmation, he was able, always in his intentions, to definitively break down Einstein's principle of locality: the measurement of a quantum entity by an observer influence, if they are part of the same system, also the other, instantaneously and at a distance of light years. It is the definitive confirmation of the *entanglement*.

But, on the other hand, this principle also provides confirmation that Bohm's theory was on the right track: the Universe would be a whole, because there is something that justifies simultaneity and that is unknown to us, although *Local Hidden Variables* would no longer have reason to exist because the results are deterministic and perfectly predicted by classical physics.

Aspect experiment, while it is true that it definitively closes the EPR case, it is also true that in a more general context and, perhaps in a less physically orthodox way, it raises a series of questions, along with other unresolved questions, about what it really is current physics and in which direction it is going.

Alain Aspect empirically found the truth that two entangled quantum objects are inextricably connected.

The correlation between the microscopic and the macro world remains unresolved.

1985 Giancarlo Ghirardi, Alberto Rimini and Tullio Weber

The question, of a disarming naivety, is still unanswered so far: why in the world around us we do not see the strange effects predicted and observed at the subatomic level?

Three Italian physicists tried to formulate an answer to this question: Giancarlo Ghirardi, Alberto Rimini and Tullio Weber. With their *GRW Theory* they have shown that the transition from a subatomic behavior to that of the visible world and beyond is probably regulated by an intrinsic property that is revealed in the progressive aggregation of many subatomic elements.

Perhaps there is nothing new: with a new name and different starting hypotheses it would almost seem like a return to reconsider a theory previously expressed, in this case by Bohm, with his *Hidden Variables*.

The *GRW Theory* is one of the many possible branches of the Physics tree that could bring unexpected fruits to *"Quantum Mechanics"*.

Another Absurd Observation: The Tunnel Effect

In the quantum world, occur events that have no comparison in Classical Physics and observers that is the scientists who carry out the experiments, often have difficulty in recognizing their real meaning.

The impossible can always happen suddenly and often there are no motivations in the knowledge gained up to that moment, which are able to make sense.

We have seen that there are entangled particles able to communicate instantly, thus breaking the barrier of the speed of light. Now we will see there are some particles that can escape from a closed place by passing through obstacles.

This phenomenon, which is called *Quantum Tunneling,* is another incredible property that occurs in subatomic world.

In classical physics, it is intuitive that a particle cannot overcome apparently impenetrable barriers because it does not have the necessary energy to do this: the simplest example to assimilate is that a bullet which, if it does not have sufficient strength, will never be able to cross a wall. Or when we play squash, the ball bounces off the wall and comes back. Instead, in the quantum world, sometimes, a particle passes through the wall. Although rare, this is a possible event and an atom or an electron can not only pass, but simply be on the other side of an obstacle.

In 1928 the Ukrainian physicist George Gamow realized for the first time that there must be something unusual to explain alpha decay[19], that is, an alpha particle is emitted from a nucleus

---

[19] Alpha decay occurs for no apparent reason to cause it, so this is an energy-generating phenomenon; and it is random because we do not have the

because it overcomes its potential barrier, and in simpler words it has crossed an obstacle which in classical physics would be insurmountable.

In this way Gamow answered many questions explaining various chemical compounds and radioactive decay.

Researchers have been asking themselves more and more questions about how this was possible and how long the particle is crossing a barrier.

The general explanation is always the same: subatomic particles satisfy the double wave-corpuscle nature. There are two possible interpretations of the phenomenon. The first is the equation developed by *Schrödinger*, which provides a possible solution in a particular case: there is a probability, though very small, that the parameters of the particle that meet the equation correspond to those of a particle that has passed through the obstacle. The second possibility is directly related to *Heisenberg's Uncertainty Principle*: not knowing the exact position of the particle favours the hypothesis that it can be anywhere and therefore also on the other side of the obstacle without breaking the barrier.

For the *Heisenberg Uncertainty Principle*, we can never observe a particle in the act of crossing the barrier, but only before and after this transition has occurred.

This theoretical part is also hard to accept, as it openly violates the common sense of the macroscopic world where the law of energy conservation applies to those involved in the work, which precludes a particle from overcoming an obstacle without

---

certainty that the decay will occur, but only the probability that it can happen without ever knowing when this will happen.

sufficient energy to do so. The bullet or the ball, infact, do not pass through the wall simply because they do not have enough energy to do so.

Those that appear to be only theoretical results are actually fully confirmed by the experimental tests and have been used in the realization of devices that we use in everyday life: some types of computer memory, EEPROM, use these technologies, as well as some microscopes or, in electronics, the tunnel effect diodes.

The interpretation of this event through the Schrödinger equation suggests that the phenomenon is not immediate, unlike the interpretation with the *Uncertainty Principle*. Only recently a team of physicists measured the duration of this phenomenon, which is very fast but not immediate.

Once again *"Quantum Physics"* surprises with its unpredictability and contradictions. In fact, scientists have taken note of what is happening, but they cannot explain why this is happening, let alone what is happening during tunnelling. It's almost like an instant transporter effect.

In 1962 Thomas Hartman, a semiconductor engineer for Texas Instruments, proposed that a barrier, regardless of its thickness, acts as a shortcut: does this mean that quantum tunnelling violates the speed limit of light?

The question is still unresolved: sixty years from Hartman's observations, regardless of the various attempts to change the points of view of the calculation of the tunnelling times or the precision reached in the laboratory to calculate it, this effect seems to be superluminous, faster than light.

The Particle Accelerators

> *The shorter the life of a particle, the higher it costs to intercept it.*
>
> *First law of physical particles*
>
> *Arthur Bloch*

What They Are and How They Work

At the beginning of the 20th century, in that period that I like to call romantic for physics, we began to investigate the structure of the atom but only with the advent of *particle accelerators* we get an idea of what it really was.

Until the first accelerator was built, scientists had used cosmic rays, high energy sub-particles that arrive on Earth from the cosmos almost at the speed of light, to conduct these experiments, going with all the equipment on top of mountains, where it was easier to intercept them.

Initially cosmic rays were used because, being made up of protons for 90%, helium for 9% and heavier components for 1%, when they hit lead atoms, as a result of the impact, they ejected other smaller particles: protons or neutrons. They came to the conclusion that the nucleus had other constituents besides those already known and, from that moment on, began the search for atomic physics.

The next step was to design and implement devices that simulated cosmic rays, called accelerators, with which they began to target and crush atoms.

These devices push particles, especially electrons, at high speeds, accelerating them at nearly the speed of light, generating high kinetic energies; when they reach high energy they push them to hit the target atoms, which, due to the impact, break down in many parts, revealing their contents. All elements resulting from the impact shall be intercepted and analysed. This provides detailed information on the components and the glues that hold them together.

Until few time ago in almost every house there was a simple accelerator, in fact, the old cathode-ray TV worked exactly with this principle: in the cinema the electrons of the cathode were accelerated and acting appropriately on electromagnets, they were directed in the vacuum of a cathode-ray tube; Then they crashed into the phosphorus molecules of a screen located behind the outer glass and broke them, causing an illuminated point called *pixel*.

The accelerators are much larger and accelerate particles with electromagnetic waves artificially generated at much higher speeds: if you want a visual example imagine a surfer pushing a wave in the chosen direction. The more energy we inject into the wave generation, the faster the particles become and the more pieces we get from the impact.

We can say that the accelerators are of two different types: the linear ones, where the particles are fired as if on a perfectly straight track and reach the target; or the circular ones, where the particles rotate along a circumference until the target is centered.

In linear accelerators the particles are accelerated in a long vacuum copper tube pushed by wave generators called *klystrons*[20]. The electromagnets make the particle beam coherent and guide it to hit the target at the end of the tunnel, where there are often interchangeable instruments made specifically to capture the events caused and to record their data. A nearly 3km long linear accelerator is located in California at the Stanford Linear Accelerator Laboratory (SLAC)

---

[20] From Wikipedia Klystron: A free-electron, linear-beam vacuum tube. From the Greek word κλύς, which is the representation of the breaking of the wave on the beach, and the suffix -tron which indicates the electronic nature of the system.

Circular and linear accelerators use the same operating principle, only in circular accelerators the particles rotate repeatedly and, at each pass, the magnetic field is amplified to the desired energy and only in that moment the target is placed to be hit inside or near the detectors. In 1929 the first accelerator was circular.

The operating schema combines all accelerators: there is a particle source that injects them into a tube where previously a thrust vacuum has been made, where they are accelerated by *Klystrons*, which operate from microwave generators which in turn produce waves on which the particles travel along the entire tube. Electro-magnets, conventional or superconducting, force the particle beam to become coherent and, if necessary, direct it towards the target where detectors, devices built on purpose, recognise the pieces and radiation resulting from the impact.

Other ancillary components are the systems that produce vacuum by removing air and dust from the tube and the cooling system, which dissipates the heat generated by magnets.

Finally, more frequently updated, there are computers and electronic systems that control the operation and analyse the data from the experiments.

Operators, technicians and the public are protected from the radiation generated by the experiments thanks to a screen and inside the accelerator there is a closed circuit monitoring system that allows you to see what happens at all times.

The power supply system plays a very important role. Finally, not directly forming part of the main ring, there are additional storage rings where unused particle bundles are temporarily stored, making them rotate continuously on a secondary rings at much lower speeds.

When the researchers started using the first accelerators, they immediately realized that they had a formidable tool in their hands to improve research, because the results immediately went well beyond best expectations: further the particles they already knew as components of the atom many other unknowns were emitted after the hits on the target and the more the beam energy increased the more particles were discovered. Many of these new particles have a very short life and are only detectable for few fractions of second, less than a billionth of second. Others particles tend to regroup and form more stable particles. The method of detection and recognition for force-carrying particles, also them released in collisions, was different.

The ability to organize and confirm the behaviour predicted by these particles, led to the *Theory of the Standard Model of Atoms*.

CERN e LHC

At the end of World War II, the scientific research leadership had moved elsewhere and Europe was out of world-class research. The French Raoul Dautry, Pierre Auger and Lew Kowarski, the Italian Edoardo Amaldi and the Danish Niels Bohr imagined a joint European effort to continue atomic research, building a laboratory and forming a group of all favourable countries.

In Lausanne on the 9th of December 1949, Louis de Broglie formulated the first proposal and in 1952, after a long series of mainly bureaucratic passages, eleven European countries signed to establish the Foundation of the European Council for Nuclear Research, better known as CERN.

Many cities offered their seat but, after a referendum among the inhabitants, Geneva was chosen because it is located in the heart of Europe, as well as for historical Swiss neutrality. Meanwhile, during the material construction of the laboratory in Geneva, the theoretical problems were addressed in Copenhagen. Imagine for one moment how happy Bohr and the other physical founders of *"Quantum Physics"* were: started from simple experiments and hypotheses all to be verified and now finding themselves with a machine so evolved they had ever imagined. I think they felt overwhelmingly proud of their creature.

In 1957, the first CERN accelerator, Synchrocyclotron (SC) 1600 MeV energy, was started. It was used up to 1990.

On 24th of November 1959 the first proton accelerator, Proton Synchrotron (PS), came into operation from 28 GeV energy: for some times became the highest energy accelerator in the world and still today performs its work without retiring.

In 1965 the French government granted permission to the laboratory to extend the research areas beyond the Swiss border and thus, in the 1971, the Super Proton Synchrotron, with its seven Km of circumference, became the first Franco-Swiss cross-border accelerator and came into operation on the 17th of June 1976, two years ahead of forecasts.

The improvements made during the design and assembly phase not only allowed the work to be finished before, but also increased the energy of the beam from the original 400 to the current 450's GeV.

Such high energies have allowed us to study the structure of protons.

Until Proton Synchrotron (PS), all the operating commands were in a control room with hundreds of switches and buttons. The transition to Super Proton Synchrotron (SPS) would have increased the control devices and, besides being really expensive, it would have become difficult to manage. So a solution was sought with new ideas, perhaps from the experiences of the scientists involved and, for the first time, touchscreen technology was developed. In 1976 the SPS at its start was completely controlled by touchscreen. From 1977 the touchscreen came on the market and subsequent developments and improvements led it to the daily use of smartphones, tablets, etc...

The 8th February 1988 completed the tunnel that would have contained the Large Electron-Positron Collider: twenty-seven kilometres of conference for the most important European engineering project before the Channel Tunnel. On the 14th of July 1989 in the Large Electron-Positron Collider, commonly called LEP, the first beam was fired. It is still today the largest electron-positron accelerator ever built. At that time the CEO of CERN was the Italian Carlo Rubbia.

In the last decade of 20th century, the world of computer science also competed in the search for different paths of development, with new computers, with more powerful processors and, above all, actively participated in the search for new ways to share data and information. The enormous amount of data produced by the Geneva's CERN, by statute, was to be made available to the entire international community of scientists throughout the world as quickly as possible, and so a solution had to be sought for its own development.

In 1989 it began to think that a specially designed network would bypass the obsolete technologies present at that time and so the World Wide Web became real.

In Christmas of 1990, after writing the code of the first browser and defining the parameters of the Web, the first website opened: *"info.cern.ch"*. It still exists today and if you search it with your browser you will be happy to relive the original experience.

In November the 2nd, unlike its distinguished predecessors, LEP was definitively retired after eleven years of honorary work to replace with its successor LHC...

The CERN tunnel with the LEP and its successor LHC
From:https://timeline.web.cern.ch/large-hadron-collider-lep-tunnel

From the beginning the tunnel for the LEP had been designed and built bigger, thinking of the subsequent developments, and so the *Large Hadron Collider* or *LHC* particle accelerator, which replaced it in the same tunnel, became the largest existing machine in the world.

In September 2008, at 10.28, the Large Hadron Collider with energy of 2 x 3.5 TeV began to produce results after a beam of protons was launched in its twenty-seven kilometres of circumference.

Unfortunately, on the 19th of September 2008, a breakdown caused extensive damage to the machine without causing damages to the people thanks to the safety systems that came into operation correctly. The accelerator was stopped for repairs, which lasted until April 30th 2009, when the last magnet was finally placed in place to complete the repair of LHC.

On the 20th of November 2009, the normal working routine, which was only reported after the machine had resumed working regularly, began again: scientists but not unforeseen.

On the 4th of July 2012 two experiments observed a new particle that showed properties consistent with the theoretical predictions for the Higgs Boson, which was definitively recognized by the scientific community on the 6th March 2013. All the characteristic elements of the boson have not yet been found: it continues to maintain some of its confidentiality by hiding at the eyes of investigators.

The growing technological innovations of LHC components allowed continuous technical updates and the 3rd June 2015 the LHC was brought to work up to the 13 TeV record energy.

Thus, an era of new discoveries opened up for the Physics of High Energies: the questions to which one seeks the answer are intriguing and fascinating for both the scientist and the profane and urge us to be interested and participate in what happens at CERN.

What gives matter its mass?

The 96% of the Universe is invisible, but what is it made of?

Why do we have more matter than antimatter?

Is it nature that prefers it or, if it really exists, is there another reason?

How did the first moments of the Universe evolve?

It is an adventure that must be shared with all open minds and that will perhaps allow us to get where no experiment has ever led us.

Even today, the LHC is the most daring scientific project ever conceived by the human mind.

But one question is legitimate: if it is true that the results of the experiments fall within a predictable field anyway, are we sure that all the predictable has been taken into account? Or is it?

By applying these considerations to research carried out in the microcosm and especially in the macrocosm, is it possible that we are missing something that we are not at all able to understand and that makes all the observations incomplete?

The results obtained from the research done in this field are found daily in our lives and they are discoveries that have proved to be winning weapons in their field of application: PET scans used to diagnose tumors or adrotherapy to cure them, to name only the latter.

At CERN, there are also experiments that lick other fields: between 2008 and 2012, a beam of neutrinos was generated and after crossed the Earth's crust, it was detected by INFN Laboratories of Gran Sasso after 2,4 milliseconds and 730 km.

Another fantastic experiment, which could profoundly affect our knowledge of physics, is the CERN Axion Solar Telescope (CAST), which looks at the center of the Sun for *axions*, likely generators of invisible dark matter: they are all still to be verified and which the scientific community hopes to answer in the next future.

CAST is a hybrid telescope born from collaboration between particle physics and astronomy.

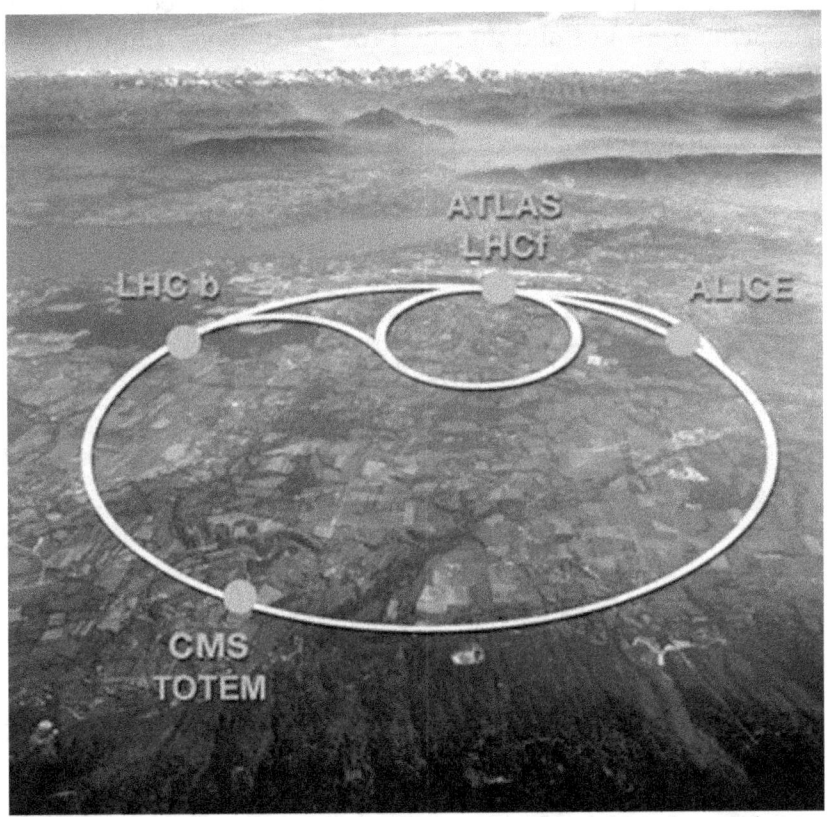

Aerial image of the LHC. The image is superimposed on the accelerator path and the location of the main detectors installed on the beam (INFN / LHC Italy)

From: La Repubblica.

Currently the countries collaborating at CERN have become twenty-three, all European nations plus Israel. Being a common scientific laboratory, it is open to the collaboration of all, for which China, Russia, the united States, Korea also participate.

On 28 September 2018, CERN inaugurated the Esplanade des Particules, a space specifically designed for pedestrians and sustainable transport and intended for the involvement of visitors and the public.

The official address of CERN becomes Esplanade des Particules, Genève.

The Last 50 Years

An expert is a man who has made all the mistakes which can be made, in a in his knowledge field.

Niels Bohr

The Standard Model

The work previously done by all physicists and researchers, from the end of 1800s had now allowed reaching a considerable knowledge and representation of the structure of matter that encompassed all the present in the Universe. It may be summed up in many combinations of few particles subjected to the four fundamental forces and at the laws of their interaction. In the period following the end of World War II, until the 1960s, physicists were overwhelmed by the discovery of a large number of new particles and began to think of something that could integrate them all together in a single vision.

Thus, in the 1970's, a theory was developed which, in its intentions, had to include all progressive discoveries. It was called *Standard Model*: this theory, in the plans of its creators, would coordinate all knowledge of the subatomic world and also possible subsequent discoveries.

The elementary particles, which are now recognized as the bricks of matter, are called *fermions* and are divided into *quarks* and *leptons*, which in turn consist of six elements each of these called *flavours*.

There are also other particles outside matter that act as mediators for forces. The *mediators* term means that the interactions between *fermions*, i.e. *leptons* and *quarks*, are possible, because there are other particles called *mediating particles of the forces*. When interactions occur between particles there is a mutual exchange of mediating particles called G*auge Bosons*, which are involved as, we can say, common friends of leptons and quarks, they deal with *mediating* this relationship in accordance with certain fixed rules.

Gauge bosons are of three types:

- *Photons,* responsible for electromagnetic force;
- *gluons,* responsible for the strong nuclear power;
- *W and Z bosons,* responsible for weak nuclear power.

The fourth Boson, which had already been dealt, was the theorized and mythical *Boson of Higgs*, for whose discovery the LHC was on purpose built.

The experiments carried out in laboratories around the world unequivocally and with great precision confirm the validity of the *Standard Model*.

So far we have seen that atoms are made up of three types of subatomic particles: electrons, protons and neutrons. Protons and neutrons, discovered in the progression of time, are not fundamental particles but in turn are made up of two types of *quarks*, which we will see in the next chapter, called *up and down*, held together by the strong force.

Particles that are combinations of quarks are called *adrons*, have a mass and we can find them in the nucleus. The two most common examples of adrons are protons and neutrons and each of them is a combination of three quarks.

The electron, which does not appear to have other components, is a type of elementary particle belonging to the category of *leptons*.

Other particles that can be classified in leptons class are the *electron*, the *electronic neutrino*, the *muon* and the *muon neutrino*, the *tau* and the *tauon neutrino*.

Electron, muon and tau have considerable electrical charge and mass.

The *neutrinos* are neutral, as their name also says, and have a small mass.

Experiments continue to search for other particles. So we have to ask ourselves a question: why, beyond this configuration, which satisfies it anyway, does the *Standard Model* need new particles?

The *Standard Model*, despite everything that physicists who use it every day may think, has not provided all the answers, as it is not a complete model. Something is still missing, and the big problem, which is not accepted, tells us that we do not know the extension of the missing part: it could be smallness, or it could be something that forces us, in order to square everything, to completely review our convictions from the foundations.

The four forces used for energy exchanges between the particles of the atomic structure serve, above all, to explain the evolution of the Universe.

Interaction type	Effects	Type of bosons used
strong	the strongest is expressed over a very short radius only at the level of the subatomic particles	Gluone
electromagnetic	infinite reach but much stronger than gravity	Photon

weak	it occurs over a very short radius only at the level of subatomic particles	W and Z
gravitational	the weakest force but with effects up to infinity	Graviton (not yet und)

Gravity is not part of the *Standard Model*. And this is already one of the reasons why the *Standard Model* is incomplete.

The *Standard Model*, on the other hand, perfectly explains how the three forces work which in the *bosons*, which carry energy and interaction information, are emitted and absorbed by the particles.

These particles are supposed to be exchanged when forces occur. A force is defined as a push or a pull. But this says nothing about how it applies: Richard Feynman assumed that forces interfere when two particles exchange a *boson*, the *gauge particle*.

Let us take an example: there are two people who on roller skates exchange a ball. Whether you throw the ball or you receive it, we feel pushed in a different direction from the starting one. The skaters of the example are the fundamental particles, the ball's throw is the vector of force and the reception is the applied force. In the case of particles, we see the force and the effect, but not the exchange: so the exchange is Gauge's boson.

In 1800, James Clerk Maxwell brought together in a single *theory of electromagnetism* all the observations, experiments and equations made up to that time, bringing them together and unifying them into the so-called *Maxwell equations*.

Using increasingly powerful accelerators and with ever higher temperatures and energies, it has been seen that *electromagnetic force* and *weak force* can be combined in *electroweak force*.

Some physicists believe that, initially, immediately after the *Big Bang*, all forces were combined into one. Einstein, after discovering relativity, continued throughout his life to search for a model that united all forces in a single unified theory, called *GUT, Great Unified Theory*.

If one day scientists can find it, perhaps we can finally understand what happened immediately after the *Big Bang* and integrate the microworld and the general theory of relativity that explains how the macrocosm works, in the only *Theory of Everything* that can explain the Universe organically, from the subatomic world to the macroscopic universe.

This is impossible at the moment because of the mathematical incompatibility of the results of the *Standard Model*.

Gravity at subatomic level has such negligible effects that it can be ignored and the *Standard Model* works well despite this exclusion.

What is missing?

The Standard Model although is the best model of the subatomic world currently available, following the new discoveries leaves many questions unanswered, one of which, the most important, is: *what is the true nature of dark matter*.

Alternatively, assuming its existence hypothetically, we must ask ourselves where the antimatter that must have been after the *Big Bang* ended up.

Other questions are still unanswered: why do *quarks* and *leptons* have such different mass values?

What is the actual role of the *Higgs boson*, a fundamental pillar of the *Standard Model*?

Perhaps the *Standard Model* is just a detail of a larger picture that includes a new vision of Physics, perhaps still hidden in the depths of the subatomic world or in the farthest corners of the Universe.

Perhaps FermiLab or the LHC, or its already designed heir, will help us discover new truths.

More trivially, perhaps, we need new eyes and new minds that are not conditioned by the memory of the past.

In addition to the Standard Model, there are other theories that try to gain credence in the eyes of scientists. With different mechanisms, but still worthy of attention.

In the *"Technicolor theories"* the role of the boson is entrusted to a set of particles hypothesized but still to be discovered and cohesive in a different type of strong interaction.

Even the *GRW theory* from the Italians Giancarlo Ghirardi, Alberto Rimini and Tullio Weber, which we have already seen, would deserve a chance.

And I also mention the suggestive theory from Anthony Garrett Lisi, based on a mathematical model that uses the Lie algebra of the E8 group, or the splendid Unified Field Theory of Heim.

Surely there will be others, unknown to me and in development, hidden in the minds of various scientists around the world.

Higgs Boson's Story

In 1964, Peter Ware Higgs (Newcastle upon Tyne, 29th May 1929), based on his calculations, provided that a basic constituent for the construction of matter, had to exist. This particle, according to its prediction, would confer the property of mass to all other particles, which in turn confer it to the whole world around us.

For these characteristics, the most enthusiastic have attributed to this particle the divine properties of creation.

The question was how to find the Higgs Boson.

There was only one possible way: to build a specially designed machine, using as a starting point, and to upgrade it, existing achievements so as to be able to identify it. This machine was the heir to the first particle accelerators.

Normally Physics, and more specifically particle physics, does not collect the media successes of politics or gossip on the most famous characters, yet Higgs Boson stole the scene by imposing himself on the attention of the general public.

Few really know what it is, but now we all think to know what it is: Higgs received the Nobel Prize in 2013, after the announcement of CERN scientists in the 4th of July 2012, because a particle appeared to correspond to the models expected.

The discovery led this branch of physicists to steal the front pages of many newspapers.

Before his discovery, this particle had become a real torment for all scientists, so much so that Leon Lederman, who also won a Nobel Prize for Physics, wrote a book on this subject in 1993 entitled *The Goddamn Particle*: *If the Universe is the Answer, What is the Question?*

During the editing the book was not accepted by the editors, who opted for a more canonical and less dangerous, according to them, *"The particle of God"*.

This title has, of course, left displeased all those who do not want an admixtures between the holy and the profane.

Surely the Higgs Boson has nothing to do with any theological proof, it is simply another missing element that serves to complete and, not even definitively, the *Standard Model*.

In addition to the Higgs Boson, physicists have come to an extreme theorization that none of the particles have intrinsic mass, but this is achieved by passing through a field known as *Higgs Field* that extends throughout the universe.

The concept that there can be a *layer* that permeates the entire Universe is therefore reproduced.

Like all other fields, Higgs also needs a carrier particle that involves the other particles: this carrier particle is the *Higgs boson*.

Quarks

The Quarks are considered elementary particles that are indivisible, fundamental building blocks of sub-atomic particles.

The term Quark is short for qu (estion) (m) ark, "question mark" and was first used in 1964 by the American physicist Murray Gell-Mann, who referred, for no apparent reason, to a passage of James Joyce's novel "Finnigans Wake".

In 1969 the physicist was awarded the Nobel Prize for his work with quarks.

In summary: atoms are made up of electrons, protons and neutrons and, these last two particles are made of quarks.

String Theory

A separate speech deserves another theory jumped to the honour of the chronicle quite recently. I draw it individually because it requires, in order to exist, the existence of the *Standard Model*.

String Theory is an amazing new way of explaining the physical world around us, modifying the established knowledge from the ground up.

The *Standard Model* becomes only the introduction to concepts that would have been impossible to understand from scratch without it.

In 1968, looking for alternative solutions to justify some controversial points of the *Standard Model*, Gabriele Veneziano, Italian theoretical physicist at CERN in Geneva, proposed *The String Theory*, a mathematical theory that attempts to find a solution to some contradictory aspects of the *Standard Model*.

The String Theory has received objections and consensus among the workers: its supporters believe this to be the possible answer to the search for the *Theory of Everything*.

It has certainly proved to be one of the most plausible, but it is also one of the most complexes, both to be described and to be intuited; so much so that in some points it seems confused.

In addition to the intrinsic weirdness of the theory, the real difficulty for scientists is to test the ideas proposed experimentally.

What we call particles, in fact, has no shape, weight or dimension: in reality they exist and cause effects, up to the constitution of the world we perceive, only because in motion.

String Theory proposes to generalize this idea of a fundamental constituent identity into one only dimension.

According to this revolutionary theory, the vision of the Universe and its elements changes completely: since so far the many types of pointed elementary particles are unable to explain the interactions of all four forces, the model is simplified by replacing them with a single entity, a string to a single dimension, very small, as the length of Planck, 10-35 m, which vibrates like strings of a musical instrument, must correspond to the different quantum states. This means that all elementary objects, which we have seen so far in the *Standard Model*, are only small strings and all elementary particles can be described as strings with different quantum states.

Thus the *Standard Model* becomes a derivative of *String Theory*.

When you arrive to the observation on a larger scale, you no longer see the strings, but the particles we know.

In the *Standard Model*, if we go backwards from an object, for example a drop of water, this is made up of molecules, in turn made up of atoms made up of electrons, protons made up of quarks and neutrons also made up of quarks that cannot be observed individually but only as constituents of adrons. To this day, they would seem to be the last solid state of matter aggregated by forces and dimensions. We replace these dimensional elementary particles with constituent one-dimensional strings with an only one dimension, the length.

These *strings* can be *opened* or *closed* and, open strings can combine in a new open string or join in a closed string: the interactions between strings generate up to five different theories on strings, which are part of the *Superstrings Theory*.

The different tension of the strings causes different vibrations and produces all forms of elementary particles.

In addition to length, width and depth, there are three other dimensions that we do not experience because they are grouped in very small spaces and the strings are so small and curved that they cannot be observed directly. It is as if we were looking from a distance at an electric cable that looks like a straight one-dimensional line but is actually cylindrical and there are still other cylinders inside it.

If all particles are identified by a vibration, even the carrying particle of the force of gravity, the graviton, falls back into the predictions and this would lead to a possible *Theory of Everything*.

String Theory has evolved into Superstring Theory.

As we have now learned from *"Quantum Physics"*, even the mathematics of string theory is not uniquely solvable and so far has not realized any predictions different from those made by other theories, but for physicists it is very fascinating.

Perhaps it would be better to dedicate yourself to Sudoku, which is certainly less expensive.

Suffice it to say that to date the mathematical structures of strings are so complex that no one even knows the equations of the formulas of this theory, but only partially solved approximations.

The now usual question arises again: we must ask ourselves if the exploration of the structure of matter has led us or put us in a point where the matter is not there and becomes something else.

Perhaps only a mathematical structure, since a wave is nothing more than an oscillation of a structure, it is therefore more a mathematical entity than a material one.

This could mean the realization of an old idea of Pythagoras: all the research of physicists was in vain because, in the end, matter is nothing else that "mathematical structures".

Karl Popper (1902-1994), Austrian philosopher argues that a theory that cannot be refuted by any experiment is not scientific.

Therefore, if string theory does not find an experiment that contradicts it, according to Popper's mindset it cannot truly be called a scientific theory.

It will be only the future that can have the last word on String Theory.

Big Bang and Particle Physics

> *"Quantum Mechanics" describes nature as absurd from the point of view of common sense. And yet it fully agrees with experiment. So I hope you can accept nature as She is: absurd.*
>
> *Richard Feynman*

With continuous accelerator improvements, the experiments have shown how everything in the universe is connected and the experiments try to find a GUT, a Great Unified Theory that aims to close the circle between *"Quantum Physics"* and *Cosmology*, with the search for a single mathematical theory that satisfies any physical phenomenon in the Universe you want to interpret.

At the present, things are much more complex: there are more levels, not rigidly determinable, which, although they have many points in common, are governed by different laws. Gravity, for example, does not significantly affect the subatomic world, but without its laws we would not be able to explain the motions of the great heavenly bodies.

We are certain of one thing: all the phenomena we describe happen in the Universe around us.

Many cosmologists indicate the date of birth of the Universe immediately after the *Big Bang* that took place 13,8 billion years ago and, since then, has been expanding at an ever-increasing rate.

It is not possible to define what the *Big Bang* was and whatever interpretation is given there is no certainty.

The *Big Bang* is an incomplete attempt to explain the evolution of the Universe from a very small and compact initial state and, of which, we ignore everything, to what we feel today.

So *Big Bang Theory* does not explain at all what the Universe could have created, or what was before it or what is outside it.

Another big misunderstanding is that there has been any kind of deflagration: it is a limitation of our understanding capacities that, similarly, we explain with what we can most understand.

For what we know, from a certain point onwards the Universe is there. That's that.

From that moment onwards, the Big Bang theory explains the expansion of the Universe.

This, despite some versions of the theory suggests an incredibly rapid expansion perhaps superluminous, but, I repeat, without any explosion as we mean it.

Big Bang theory, like *"Quantum Theory"*, also expresses concepts that contradict the way we perceive the world.

In the early stages of the Big Bang it is assumed that all forces, initially concentrated in one unified, have separated.

The more we can study what happened in the early stages, as if we were in a time machine that takes us back and allows us to see when the Universe was still just born, the more we realize that no scientific theory can be formulated, because science itself cannot apply and our certainties collapse.

The first formulation of this idea should go back to Georges Lemaître in 1931.

Initially, according to the theory, there had to be unimaginable densities and temperatures that produced energy.

Since then and for reasons currently unknown, the Universe has expanded, becoming less dense, cooling.

During the expansion matter was formed and radiation began to radiate dissipating energy in a few moments. Thus, from a singularity, the Universe was born.

I would like to make a hypothesis from a concrete example: let us take an old incandescent light bulb, made up of a filament of a material resistant to high temperatures and from a pool-sealed bulb in which the vacuum was previously made, so that there is no oxygen that can be burned by the heat of the filament. After turning it on, it will begin to spread light in the room in all directions. This is because the filament, crossed by electricity, becomes incandescent and the electrons that make it up, excited, are accelerated in all directions emitting light. The duration of the lamp can thus be very long.

Let us now move on to the second phase of the experiment: let us now practice a tiny trout in the bulb and instead of the pre-existing vacuum there will be the same air as the surrounding environment.

Let us give the lamp to someone who has never seen it work before and who does not know its duration in advance: however, the lamp will light up, with less intensity and for a shorter time. The observer will see something magical that produces a light from nothing and that will work until, due to the presence of oxygen, the filament is damaged and no more luminosity is produced.

For an observer who only sees the light bulb with the trout and describes it, this would still be a kind of magic that for a short period has illuminated an environment.

Having no knowledge of the ideal operation, the observer will never know that the light bulb was missing anything.

The most complex thing to understand and perceive is that, since the missing part is the vacuum, it would not be detectable at the back and the observer could never imagine that there must be another component that allows the lamp to function ideally.

On the basis of this example, and referring to cosmological studies of the Universe and its elements and possible reunification theories, I have doubts that our research lacks precisely the identification of something of which there is simply no trace, such as the void of the bulb and of which we have no perception of existence.

The initial point is incomprehensible to us because we lack the terms and knowledge to define it. At that moment matter, energy and space were concentrated and compressed in a volume equal to zero and had nothing to do with our common idea of point. The whole of this, which we cannot understand and define, is called *singularity* by cosmologists.

Many predictions of this theory have been confirmed by the observations and for this reason it is the most accredited theory on the development of our Universe.

It is clear that, because of the limitations of our knowledge reflected on our formulation of the laws of science, we do not have the appropriate instruments, either scientific or mathematical, to formulate hypotheses of any kind on the moment the Universe was born. On the contrary, the next moments are now within our grasp.

To this day we have managed to go back to a time of 10^{-43} seconds immediately after the birth of the Universe. In other words, we take number one and move the decimal point to the left for 43 times.

In the latest theories to explain these initial moments, we began to talk about *Quantum Cosmology*. For Cambridge University cosmologists, in the early moments of *Big Bang*, the universe was so small that, in it, there were no laws of classical physics. But *"Quantum Physics"* already had possible explanations. Scientists are now trying to understand how and when the laws of classical physics came into being in later times.

Perhaps, according to the latest calculations, at 10^{-43} seconds, the Universe was so small that it covered an area of only 1×10^{-33} centimeters.

The expansion in the immediate aftermath led it to expand to an extension of billions of light years. At this stage, according to the proponents of the *Big Bang theory*, matter and energy were all one, just as *the four fundamental forces were a single force*.

Based on the projections made, the temperature at this expansion stage was 1×10^{32} Celsius.

All the expansion that led to the formation of gas clouds should have taken place in less than billions of seconds and cosmologists called it *inflation*.

In the expansive phase the Universe has cooled and at 10^{-35} seconds and, in this phase, matter and energy have separated.

Bariogenesis has begun, i.e. the birth of *barionic matter*, which is not matter as we understand it, but are the attempts of the first aggregations of non-elementary subatomic particle, consisting of an odd number of *quarks*, the *barions*. The *barions* belong to the *adron family* and take part in the strong interaction and will become the type of matter that we can observe.

There is another type of matter which has been hypothesized, *dark matter*, which we cannot observe, but we know it exists because it affects energy and all other matter. It is believed that during bariogenesis in the Universe there were almost the same quantities of matter and antimatter. A small part of the excess matter survived, while all the rest of matter and antimatter annihilated each other. The remaining particles would then have combined to form all matter that exists and is visible in the Universe today. This would also explain why it is so distributed in such an enormously empty space: because matter is scarce.

At 10^{-11} seconds from the beginning of everything begins a cosmology of particles, but above all we have eyes watching what happens.

Our eyes: with the current accelerators we can simulate in the laboratory the dynamics of the particles of those moments.

The data we obtain tells us, even if partially, what the Universe was like at those moments.

The unified force was fragmented and the Universe was so dense that the light of photons, more numerous than particles of matter, could not illuminate it.

At the end of 0,01 seconds after the initial moment, comes the period of standard cosmology and our knowledge becomes very profound. Scientists now believe they have very reliable data on what happened.

In particular, as the Universe continues to expand and cool, neutrons and protons appear.

At one second from the initial moment, these particles have thickened in the first nuclei of light elements such as deuterium hydrogen, helium and lithium.

The process is known as a *core synthesis nucleus* and takes place in a universe still too dense and to hot for them for existing stable atoms.

All of this should have happened in a single second but, as for every route, however long it may be, you always start with the first step.

After a hundred seconds, the temperature drops to 1 billion Celsius and the subatomic particles continue to combine and distribute at 75% in hydrogen nuclei, at 24% in helium nuclei and the rest in light elements, nuclei without electrons because the temperature was still too high.

From this moment on a tourbillon of cosmic events has led to the formation of atoms, stars, galaxies, planets and in general to matter with the great contribution of gravity force, now ubiquitous.

Finally, after about 300 million years from the initial event and about 5 billion years ago, our solar system appeared.

Currently the temperature of the Universe has dropped to -270 Celsius very close to absolute zero and it would appear that the Universe continues to expand, but we cannot say if this trend will continue forever.

According to the *Theory of General Relativity*, this depends on the quantity of matter and the force of gravity.

One hypothesis is that after an expansion it can contract again until it collapses and then start again in an infinite cycle, to expand again. This bounce theory is called *Big Crunch*.

For the Big Bang, the center of the universe does not exist.

Many cosmologists are critical of this theory, because it violates the first law of thermodynamics, it also violates the law of entropy, and many interpretations of the measures taken, such as the move towards infrared and the measurement of background radiation, are not correct.

There are alternative models and others will continue to be proposed, but the charme and resonance of the *Big Bang* will remain unchanged forever.

2021 Has Physics Lost the Right Road?

No remorse for atoms: they can experience all possible experiences at once. They really follow the advice of the great baseball player Yogi Berra: "If you find a fork in the road, take it".

Jim Al-Khalili

Madness or a New Path: Double Slit Experiment

The experiment of the *"double slit"* for the peculiarity of the results it offers is commonly called *"The most beautiful experiment that has ever been performed in physics"*. It uses a very simple instrumentation with results visible to the naked eye and reveals to the common observer the evidence of *"Quantum Theory"*. It can be seen how, by changing the observation method, the results we obtain, in some cases, become contradictory to those obtained previously.

We point a beam of light at a screen with two parallel slits. The passing light stops on a later screen.

The wavy nature of the light beam, passing through the slits, generates on the next screen a wave interference that produces a series of alternating light and dark strips as if each particle passed through both slits simultaneously. This experiment, which shows the behavior of waves, is more than 200 years old and dates back to long before *"Quantum Theory"* was formulated.

We do the same experiment using electrons, sending one at a time through the slits. Along its path there is no element of disturbance and, having to do with particles, we should have a "sum" distribution at the passage for each individual slit. Electrons should be distributed mainly behind the slit through which they have passed; instead, there are still interference bands equal to the waves previously seen, as if each particle passes simultaneously through both slits interfering with itself. This phenomenon of contemporary is called *"overlapping state"*: an electron is as if it had a corresponding of itself with which it interferes, an electron starts and we reach the image of a double electron.

But there is another event definitely unusual.

Leaving aside the theoretical explanation that would require very thorough knowledge of Mathematics and Theoretical Physics, we will confine ourselves to the references provided by Bell's theory, which has been possible to experiment, even with many "escapades", only from the 1980s, because of the instrumental difficulties.

Once a detector is placed inside and behind one of the slits, we check the passage of a particle and see that the interference disappears. This means that than just observing, without any intervention disturbing the particle, changes the expected result.

Pascual Jordan, a physical colleague of Niels Bohr in Copenhagen, said:

> *observations not only disturb what has to be measured, they produce it... We force [a quantum particle] to assume a definite position."*

This means, in simple terms, that with our only presence, we determine the results of the measurements.

> *In other words, Jordan said, "we ourselves produce the results of measurements".*

Then we could bluff with nature: we detect the particle after it has passed the slits and so we will know which path it has "decided" to choose.

John Wheeler, an American physicist, devised an experiment on this "delayed choice" which was actually carried out in the 1980s. The verification is amazing: the result does not change. The particle seems able to predict our observation and behaves, only in that case, as if the observation has been made: it foresees our intent to observe it.

It is as if the particle, leaving the state of probability cloud, "collapses" into a single well-defined point and the delayed choice experiment indicates that any intent causes it to collapse. But then does this mean that the collapse occurs after we have made the choice to do so, consciously?

Do you know what this suggests? That consciousness and *"Quantum Mechanics"* are perhaps related.

Eugene Wigner, a Hungarian physicist, anticipated these conclusions in 1930:

> *It follows that the quantum description of objects is influenced by the impressions that enter my consciousness.*
>
> *Solipsism can be logically consistent with current "Quantum Mechanics".*

According to Wheeler we can come to the hypothesis that the single idea of thought modifies multiple pasts in a single present, "collapse" and, in a past and future history. We are the ones who "determine the Universe".

To date, no one is yet able to take a position regarding these experiments which leave us truly amazed; it is certain, however, that the consideration that consciousness and *"Quantum Mechanics"* are somehow connected cannot be avoided.

Roger Penrose, a British mathematician, physicist and cosmologist, began studying this connection in the 1980s, suggesting an inverse way, namely that *"Quantum Mechanics"* is involved in consciousness and not vice versa. He wondered if there are no structures of molecules in our brains that change their state based on an external quantum event and determine our response which we think is conscious and spontaneous but is actually induced.

If such a thing were possible we could also have superposition states that would cause the spark of conscious thought in the neurons which then becomes an electrical signal.

Perhaps genius could be explained just as the ability to simultaneously have normally incompatible states as a consequence of true quantum effects.

This would also explain the difficulty of an artificial intelligence to come even remotely close to the capabilities of a human brain.

We are far beyond the quantum computer: we are directly, deep down, *"Quantum Mechanics"*.

The Quantum World

On the end of the 1800s the scientists were sure that they had found the answers for all questions that physics could ask, and so there was nothing left but to improve the accuracy of the experimental results already achieved.

But beneath the ashes there was a sacred fire that, as it unfolded, would ignite the old ways of knowing to leave the place to a revolutionary Physics: the *quantum* and *relativity*, with their ingenious ideas, would illuminate much of the 20th century.

The first years of the 21st century, on the other hand, presented the opposite scenario: the questions posed by these innovative ideas proved to be much more complex and numerous than the answers that had been given and, indeed, were so many.

The research method showed substantial differences, especially in the experimental method and the equipment used: by few scientists, pioneers with limited resources and simple tools, often only paper and pen, but who managed to assert themselves thanks to obvious intellectual abilities, we have moved on to a diverse scientific community of physicists, mathematicians, logicians, philosophers, whose ideas are difficult to decipher and synthesize because, in many cases, there is no longer an experimental verification, possible and immediate: the results obtained from the various experiments need to be studied and interpreted long before any confirmation can be obtained.

In any case, it is often not possible to establish the validity of the result consistently with the proposed theory, and in all this confusion it has become very difficult to understand which theory is true and which is not.

In fact, despite increasingly costly and complex investments, policies and experiments, carried out in very sophisticated underground laboratories similar to giant starships, up to the observations and tests carried out on the Space Station, we have not yet obtained certain experimental checks and were able to dispel the residual doubts; in addition to aggravating the point of view of a neutral scientist, the lack of a verification was counted not as a missing result, but as a new starting point to be verified and valid for other theories, precisely for this reason, not validated or validable.

Referring only to CERN, because I am is in Europe and close to me, it is far beyond the normal scale of everyday reality: a 13000 tons machine that reaches during experiments almost 0.9999999991 speed of light; non-detectable void levels in the Universe; from temperatures that reach the -271 Celsius, therefore very close to -273,15 of absolute zero, up to over one hundred thousand times higher than that of the Sun.

The physics we came produced this technological monster. And there are many who are asking for a further step forward to be taken in the immediate to arrive at big machinery four times this one.

The purpose of all this was to research the basic components of matter, the so-called elementary particles. *The Physics of Elementary Particles* is also known as *High Energy Physics*.

Unfortunately, it is an extremely costly science both from an economic point of view and for the human resources it needs, so much so that it is increasingly called *Big Science*: many scientists, many technicians and so much money.

And it becomes more and more expensive as the size of the accelerators increases to reach ever higher energy levels.

The other objection, which many people are beginning to make, is whether this is the right way forward.

The Crossroads

Are the objectives we would like to achieve on this path, or have we simply taken the wrong direction?

We go by order.

In this way, many might object, citing as important results achieved, the Higgs boson of 2012 or the gravitational waves of 2016: these results are objectives achieved on theories expressed long before.

These are only experimental tests of theories clarified and confirmed by other related phenomena, obtained with instruments that were finally possible to construct.

The future that physicists expect is the heir of the LHC, which should be the *Future Circular Collider, FCC* and, according to the forecasts, should have a circumference of 100 kilometres, almost four times the length of the current LHC, and three coexisting accelerators and generate collisions up to 100 TeV, almost ten times the current ones.

The project promoters have outlined the new accelerator which, as requested, could be completed by the 2040.

The goal to be achieved will be to identify unknown processes and mechanisms beyond the *Standard Model*. Another possible objective will be the study of dark matter and the understanding of why matter is present in greater quantities than antimatter.

And here we come back to the sore points: with a projected cost of approximately nine billion euros, the project needs the participation of many universities and research institutes and, above all, of industrial sponsors, since the potential benefits would be equally shared by all.

But there are strong doubts about the opportunity of this expenditure: the theoretical physics Sabine Hossenfelder, who works at the Institute of Advanced Studies in Frankfurt, has also moved through her blog many critics to the realization of such a machine, Considering the necessary expenditure inconsistent with the possible results: she believes there is no possible intuition that invites you to dive into this adventure, because it is not certain that the mysteries of dark matter are accessible to this new project.

Moreover, also United States renounced its *Super Collider*, *SSC*, considered indispensable at first, which, with its 84-Km roundabout, would cost "only" 6.000 million dollars and should have been ready within ten years.

Realistically there are now two currents of thought.

On the one hand, there are those who, in large numbers, continue to support the philosophy of *Big Science*: ever greater, ever more powerful and ever more untenable ideas, in order to continue to have as the only results only likely approximations of the truth, since, for the very admission of physicists and theoretical and probabilistic assumptions, any possibility of obtaining a certain result is excluded. On the other hand, there is a faction that, based on its own flow of information, claims that Physics has become a nuisance and perhaps out off the rails of common sense.

Already with existing accelerators the number of elementary particles has grown so much that it undermines the same idea that most of them could really be considered elementary.

The discovered *particles, pions and muons* of various kinds, those called *W or Z*, not to mention the corresponding antiparticles, have reached such a large number, let us speak of hundreds, that scientists have begun to talk about a *particle zoo*, a zoo with too many occupants.

Now the terms for defining them, as colour, flavour, up, down and so on, are increasingly difficult to find to describe the logic and hardly respect the meaning that we associate with these concepts on a daily basis.

Scientific journalist John Horgan, in the 1996, also claimed that the search for other elements of the subject was reaching a point beyond which there would be no further progress.

The starting question, if this path was the right one or simply the direction taken was wrong, perhaps offers another point of view.

Are There Other Ways?

Is there a third way?: if physics really became affected then there must have been a time when it was allowed to happen and, the proponents of a research that must progress at all costs, whatever the direction taken, should take responsibility. of a lack of certain answers to the questions that had been asked: physicists must not be dictators of "everything is possible, if it pleases my way of seeing", but change the point of view in "everything is possible, because I study and verify the proposals of my competitors or my colleagues, and I am willing to abandon my view, because someone else proposed a better solution". It is simply called humility.

Other disciplines have developed methods to understand if there was progress or not: for example in *Computer Science* there was Moore's law[21] which, through a simple relationship between electronic components and time, is able to establish whether a technological progress grows up consistently with forecasts. Perhaps something similar should be sought for Physics, which validates in the aftermath a path taken by the results it arrives at, or signals whether it has become just a mental exercise that does not produce useful fruits.

The objection is obvious: we cannot know if we will get results until we don't verify the theory.

We can, however, respond to this objection by recalling what Popper said and introducing a new way of seeing things: if we fail

---

[21] Legge di Moore: "The complexity of a microcircuit, measured for example by the number of transistors per chip, doubles every 18 months (and therefore quadruples every 3 years)"

to verify a theory and, even more, we fail to verify it in a suitable time, either the theory is wrong or we don't have the skills to confirm it. Consequently, as already seen for *String Theory*, a theory that cannot be refuted by any experiment could not be scientific.

In both cases we should look for other ways without getting stubborn in this "therapeutic persistence on theory".

A Possible Solution

In the past century, the experimental confirmations of *Theory of Relativity*, *Planck's Hypotheses*, *Bohr's Models*, *etc.* on average they took measurable times over a decade to be confirmed...

Today we could be more generous: we could even reach twenty years, but in the last seventy years we have gone much further.

The new ideas, such as *String Theory*, continue to have no theoretical and experimental veracity: they are only and exclusively a beautiful exercise for few theorists and this really makes us think that Physics, indeed, the physicists who establish the lines of research have gotten into a dead end.

In fact, from 1970 up today, we have not a single new intuition that has led to a certain result or even just a possible prediction. Objectively, it is an embarrassing situation, especially when compared to the results obtained in the first half of the twentieth century.

This observation leads to a question: is physical research, which should provide confirmation of theories, still in the hands of physicists, or this extreme research has so greedy everyone with its large public and private capital to invest, that it has now made the scientific research a bottomless pit channel funds where to convey and perhaps be able to express, through this system, a power torn from capacity?

It would then be confirmed that some scientists have become bureaucrats, more or less consciously, and have given the direction of research to technocrats, soldiers, politicians, sponsors, thus becoming victims of the "do ut des" policy.

The poisoned fruits of an "unreal" Physics began to be seen from 1970, the year of the presentation of string theory, but for some even earlier, perhaps from the 1960s and the model theorized by Bohm.

It is since then that scientists, conquered by the micro to the macro cosmos, have allowed themselves to be fascinated by mathematical theories that push them to seek criteria of unification and symmetry, elegance and naturalness.

And they forgot Einstein's beautiful statement:

> *If you are out to describe the truth, leave elegance to the tailor and cobbler.*

Today physics talks about supersymmetry, dark matter and energy, multiverse, inflation, strings, but how many are those who can say they know how to master these concepts enough to be able to explain them comprehensively?

The risks

The futurologist Robert Jungk in his book *"The big machine"*, speaking of the CERN accelerators, clearly expresses these concepts:

> *the more grandiose and expensive the projects become, the more they depend on factors extraneous to science itself.*

In addition to depending on powers unrelated to science, there is another equally dangerous and subtle risk: the power to administer a *"religion"*.

The scientist, as the repository of a knowledge not perceptible to the majority of ordinary people, develops the intimate conviction of being, along with his other peers, the reference of a power that is exercised on the fear that arises in all of us when we fail to understand something, a bit like it happened in the Middle Ages when alchemists were seen as magicians.

And this *metaphysical* power has its places of worship in places where to carry out increasingly sophisticated experiments, tools are needed so complex and ultra-normal as to arouse sensory emotions in the viewer, just like a religious or magical myth.

It's not a casuality that a physicist, the German Helmut Faissner, speaking with Robert Jungk during a visit to CERN, expressed these doubts to him:

> *... sometimes, I wonder in all seriousness if what we do here at*

> *CERN does not represent some kind of 'cult'. - A cult means in the language of our age that we call 'scientific' - a search, an aspiration towards something greater, towards the truth if you want to call it that.*
>
> *Perhaps this search will never cease...*
>
> *Thus we humans will be able to gain more and more knowledge, in some hidden fields, without ever reaching an 'end'.*
>
> *And so, are we not doing what men of earlier ages tried to accomplish in another world by elevating churches to their highest aspiration?*

Doesn't Jungk implicitly draw from this the belief that research can be a modern form of prayer?

On the other hand, however, perhaps even more physicists believe that this panorama is not excessive and continue to have faith in the method used so far.

Their idea of searching at any cost is similar to shooting in any direction with the belief that sooner or later, a target will be hit.

It is no coincidence that many discoveries occurred when people were looking for something else *(remember Serendipity?)*.

If this were true, even partially, it should be quietly pointed out that the money that is spent is the billions of taxpayers.

Money that is spent on a Mathematical Physics which, according to many researchers, now seems to be perched on theories that few can understand and those few, often, do not do research but they do academy.

The Education System

Richard Feyman, once questioned on the methods of university studies, commented by stating that

> You cannot get educated by this self-propagating system in which people study to pass exams, and teach others to pass exams, but nobody knows anything.

Some time ago I was outside a classroom of the Faculty of Physics and I was listening to a lesson in theoretical physics: the difficulty of the professor in explaining clearly to the students the concepts he was trying to express was obvious.

Now I ask myself a question: which student will come out of an exam taken with that professor and with the concepts he expressed during the lessons?

I will never stop remembering an anecdote I came across in the early years of my physics studies:

> *a professor entered the classroom and began to explain Theory of Relativity. At the end he asked the students if they had understood it. Receiving a negative answer, the next lesson explained it again. And again the students did not understand it. After several attempts at explanation and understanding, at the last lesson,*

> *before asking the students if they had understood it, the professor exclaimed "Ah, here it is. Now I got it".*

Which on closer inspection summarizes in itself the way of conceiving the Physics of someone like Feynman.

I therefore ask again the question asked in the previous paragraph in a different form: with this large multitude of new concepts and theories, is the person who teaches able to transfer knowledge?

It is no coincidence that the Nobel laureate Richard Feynman had already sensed the dangers inherent in such a system and which he expresses with his observation:

> *I didn't see what a self-reproducing system is for in which you pass exams to teach others to pass exams, without anyone ever learning anything*

The expression is indicative of a self-referential system that teaches very little.

Among the thoughts attributed to Albert Einstein this is often also attributed:

> *each of us is a genius but if you judge a fish by his ability to climb a tree, he will live his whole life believing he is a fool.*

I am not sure that this phrase was really expressed by the great genius, but even if it were not, the intrinsic validity is formidable.

The great hope is that every human being is potentially the one who could represent the turning point towards a different future.

Every human is a hope.

Every human we lose is a hope that cannot be verified.

Are we so rich that we can disperse such a capital so lightly?

And all those hope that, in any case and often even with difficulty, reach the possibility of expressing themselves, are they enabled to do so?

Is our school system able to support and recognize any wealth it encounters?

Even from personal experience, I really think not.

With a few exceptions, and I refer as an example to the Princeton IAS structured by some enlightened mind, this system not only requires fish to climb trees, but also that with an evolutionary leap they transform into runners without legs but only with fins.

I would like to ask many teachers: "How many students were treated as if they were that fish?"

Forced to swim against the tide in classrooms that did not exalt their talents, if they were in possession of them and, instead, forced to believe to be stupid.

To believe they are useless.

The post-primary education systems, with their teachings in preparation for any other order of education, with very few exceptions, are ancient and obsolete.

The current teaching method suppresses creativity, originality and intellectual autonomy, because, just as Feynman said, it is a system that reproduces the method of someone who studied to pass the exam and not to develop new goals.

The proof is under our eyes every day: look at the smartphone you just bought and compare it with the mobile phone of twenty years ago. Do you see the difference? Remarkable, right?

The same goes for a car: look at the cars that run in daily traffic and compare them with those of the early 1900s. It seems that two different civilizations designed them.

And now make a comparison with today's school method and that of the early 1900s: are there the same substantial changes? No. Nothing really significant has changed in over a hundred years. It is as if students are being prepared to perpetuate a past that no longer exists in the rest of society.

Proof? Just that statistic that is so loved by physics.

At the beginning of the 20th century, there were at most a thousand scientists who produced an impressive number of geniuses: see the photo of the Solvay Congress of 1927, if in doubt.

Today, with hundreds of thousands of scientists, the incidence of genes and the possibility that there are new ones is close to zero.

Statistics, inexorably, condemn the system.

The big trouble is that you will find few enlightened people willing to admit it.

No two intellects are alike but students are treated as if they are.

A repetitive method that must be the same for everyone, which must level rather than enhance skills: it is very serious and unjustifiable that levelling must be downwards, sacrificing any excellences.

A doctor who prescribes the same treatment for all his patients would cause incalculable harm. Many patients would fall ill and eventually die from it.

At school, exactly this happens: the teacher administers the same method to numerous children, each one different from the other, each with their own personal ability and learning needs.

Someone should answer for all of this and someone should feel guilty.

No, It Is Not True

The current director of the Princeton IAS is the Dutch Robbert Dijkgraaf (January 24, 1960-): he is a theoretical physicist expert in string theory.

In 2020, in one of his articles, he openly disputes the idea of physics that has come to a standstill.

In his point of view, knowing from the latest data that 95% of the Universe is missing from our cosmological knowledge, he is convinced that new challenges on the constitution of these parts, dark matter and dark energy above all, are frontiers of a new Physics that does not it will be complete until we have full understanding and this will become the trump card as the new laws yet to be found could *emerge* from a physics " composition" of many parts.

In this scenario, physics will always remain self-referential and autonomous and will respond always and only to itself, because it uses its own laws to produce new hypotheses to be reinterpreted with a mathematics that is always ready to adapt or to force, a bit like the fantastic Baron Munchausen that escapes the quicksand by pulling itself from the hair.

A New Scientific Policy

There is another doubt that systematically assails me: let's go back to CERN and its LHC accelerator for a moment.

Everything is built with taxpayers' funds, but the people who participate in these experiments are chosen from within groups that already know each other and participate in that project.

Intervening from the outside to verify different hypotheses formulated outside those circles is impossible.

So, is it right to grant the command of something public to a small circle of privileged people who in some cases insist on doing the same research systematically in order to irremediably obtain the same results?

Attention, not repetitive results, but results that always pursue the same idea?

A solution could be represented by a cyclic reset of the command board, for example ten years to give them the time to complete the phases of a research.

At the End

In conclusion, even if it is true that there has been an evident slowdown in the evolution of Physics in recent years, it is also true that we cannot ignore that this discipline, and the way it is viewed by the rest of the world, has changed so much that he is now in a position of defense even from his own players.

Identify the moment in which a self-referential Physics began to manifest, is crucial.

Where, then, will be the beginning of this difficulty?

I don't think the mechanisms that led to the formation of the cosmos have foreseen solutions through a series of equations that produce other probability functions that can only be solved only with specific mathematical operators.

Taking a long step back and returning to Planck for a moment, we recall the same constant that he did not want to present, because he believed it was a mathematical trick to make ends meet. Here, that could be a good starting point, knowing a priori the results obtained along this path.

Many report that Einstein repeatedly invoked, for the laws of nature, greater simplicity and that even Occam, as already said: long ago expressed himself in the same sense when he affirmed that, "... a thing must not be multiplied beyond necessity, and hence the simplest of several hypotheses is always the best in accounting for unexplained facts"..

Perhaps the future and the very hope of survival of physics are precisely in this.

Or the resistance we oppose to changes is just the same laziness due to the realization that after all we have become creatures of habit unable to rise again for a series of reasons induced both by the goals achieved and by our new way of life: we repeat the same gestures, don't we generally love innovators and subconsciously rebel against change?

Frank Wilczek, a MIT physicist and Nobel Prize winner in 2004, says:

> *We hope that the laws we find will be beautiful*
>
> *We hope they show symmetry and explain a lot of things in terms of a few assumptions, so that we get more results than we would expect.*

Wilczek, however, warns that there is no guarantee that this will happen, agreeing on the opinion that physics has been lackluster in recent years, because, according to his point of view, there has been

> *such a spectacular success before, in the 70s and 80s that it has resulted in such good world models that it will be very, very difficult to improve them.*

Neil Geoffrey Turok, a South African physicist, one of the greatest experts in string theory, counteracts this with a message of hope.

Turok remains optimistic as he believes that a new era of physics could be just around the corner: a different physics, which would produce bold new ideas directly attributable to *"Quantum Theory"* and *Relativity Theory*.

> *What we need is for some young people to probably come forward and say: 'Aha, this is the way it all fits together'.*

... *Nothing Has Changed: We Are Only Men!*

Errors we do using insufficient Data are Much Less than Those using no Data at All.

Charles Babbage

In 1964 Feynman gave a series of lectures to non-specialists at Cornell University on the method of studying nature used by physicists.

The scientist, who among other things was also possible to meet in Los Angeles pubs at night while playing his bongo, was not an "orthodox" physicist and, precisely for this reason, he was not welcomed by many of his colleagues.

In one of these lectures, the one on nature from the point of view of *"Quantum Mechanics"*, he warned listeners:

> *... And so don't take this lesson too seriously ..., just relax and enjoy it. I will tell you how nature behaves. If you just admit that it behaves this way, you will find it enchanting and wonderful. If you can, try not to ask yourself "But how can this be?" you would enter a blind alley from which no one has yet emerged.*

This is the honesty required of a scientist: Feynman, who also won a Nobel Prize, explained that in some cases it is also legitimate to renounce the use of mathematical formalism in the name of better comprehensibility, without the treatment losing its rigor and in effectiveness.

There is also another much more subtle aspect that in some cases makes the new physical theories incomprehensible: many of the experimental effects that we fail to understand, especially of the macrocosm but also of *"Quantum Physics"*, as well as an intrinsic bizarre to the inability to see beyond the limits of our understanding and, when this occurs, we invent completely elusive mathematical models, thus generating theories that also seem abstruse.

In reality it would take a healthy self-criticism and start from the observation that we do not know the reason for some things that happen.

The role of Physics remains today at the center of lively debates: it has achieved in depth knowledge on the mechanisms that regulate the natural world and has allowed the development of technologies for the use of its discoveries for the benefit of man, but it is equally alive a widespread fear of improper use of physical tools.

Otherwise, the consequences would be apocalyptic and irremediable: the very existence of humanity is at stake, which has its own destiny in its hands as never before.

The atomic bomb is not the only possible poisoned fruit: we must ensure that we work to avert other dangers that are still unknown.

The twentieth century also left us another legacy: it highlighted the value of excellence. It is clear that "the king is naked": in this branch there is a need for a class of scholars who are on average champions with peaks of authentic genius.

This can only arise with a teaching method that favors and promotes excellence, probably with a system different from the current one and that enables the teacher to recognize these minds and enable them to express the best.

We must recover a concept of the evolution of humanity: we are a single interdependent thing.

We save ourselves and progress all together and, just as we always put the car or the strongest man in our head, we must also be convinced to always focus on these excellences.

Side Effects of "Quantum Theory"...

> *Newspapers once wrote that only twelve men in the world were able to understand the Theory of Relativity. I don't think that's true. Maybe there was a moment when a lonely man understood something, because he was the only one who was thinking about it, before writing his article. But after its publication, the theory was somehow understood by many people, certainly more than a dozen people.*
>
> *Instead I think I can safely say that no one understands "Quantum Mechanics".*

Richard Feynman, in The Elegant Universe, from Brian Greene

"Quantum Mechanics", as well as being strange and mathematically complex, forced us to review the understanding of reality and the logical mental processes we were used to from our beliefs and habits, becoming a source of inspiration in all fields, especially for the free thinkers who they drew heavily, hoping for answers to questions that have their roots in the mists of time.

The delocalization and probabilistic theory, the ability of the electron and other subatomic elements to be everywhere but never to be perceptible, rising to the state of clouds and illusory existence, have inspired comparisons especially with the unpredictability of destiny.

Case, Coincidence, Synchronicity, Serendipity

In scientific discoveries we need application, preparation, above all ingenuity but also a lot of luck, for many the "fluke": on many occasions scientists have found a treasure looking completely different, or even not looking for it. The Curie spouses are one of the most striking examples.

The Greeks had embodied it in "Tuche" which the Romans later called "Fortuna", in the most well-known assumption to date. Originally, the term "luck" (Tiche) was derived from τυγχάνω, "happening", which in addition to randomness gave a sense of inevitability.

A good question is: do things happen by "fluke" or why is it "inevitable" that this happens? In the same question try to replace the word "things" with "discoveries".

The Greek philosopher Aristotle had the same doubts and tried to understand how the fluke, which for him was the tangle of effects caused by two free will, influence nature and our choices. For example, I decide to go to the cinema and a friend of mine decides the same thing himself. It would appear that the meeting was due to the fluke. Or was it inevitable for some hidden reason?

In the Universe, as understood by Physics, there are no independent causalities: every trajectory, every orbit, every particle follows an order inherent naturally in things and predetermined by the origins of events. If we had a computer with infinite power we could trace the motion of every element of the Universe from its origins up today and predict its future.

Everything follows a natural order of things, even collisions or events that seem to be random.

Accidental events are an evasive way of concealing our inability to comprehend a total picture.

The Universe and its components, as well explained by Newton's laws, have always been in correlation with each other and all the movements of the stars are mathematically interpretable in a *single dynamic system*.

Even the Universe that surrounds us, the cosmological and the quantum universes are all related by laws that bind the individual elements.

When we talk about an "isolated" system, this does not exist in nature, it is a convention that allows scientists to obtain a vision of the phenomenon without having to consider too many variables, induced by other nearby systems that would make the calculations too complex.

For example, while I play soccer with my friends and shoot at football door, I do not calculate the influence that the moon and other celestial bodies exert on the trajectory of the ball, yet these influences are there.

Conversely, if I need to send a spaceship from Earth to Jupiter, I am forced to calculate a trajectory that takes into account the position of all the planets involved in the journey, to avoid its attraction or to exploit it for my purposes.

In nature, all systems, although conventionally isolated, are interconnected in causality and not in randomness.

Mathematics applied to Calculus Science, then implemented on computers and applied in everyday use, generates random numbers of vital importance for many applications that we ignore: video games, applications that require encryption, for example when you send your emails, for the generation of commercial transaction passwords or pins, or simply when you want to choose numbers to randomly bet on.

In fact, this is not true because of the impossibility of generating random numbers in any programming language: there is always an algorithm that determines the extraction mode.

Let us remember that any algorithm is a set of instructions, which, according to a set of conditions, determines an output data. This sequence is well determined by the instructions and cannot produce a random result. The complexity of the algorithms increases the time needed to calculate the random number resulting in the need for ever greater calculation capacity.

The algorithms used are never really random number generators, rather they generate numbers which, however, at the end of a cycle, must always begin to reappear.

We can greatly lengthen the series, so much so that it seems random, but in the end the sequence will always and in any case start again.

This is why in spy or action movies it is always possible to trace a Pin or Password of a person or a safe: it is enough to have an increasingly powerful computer that calculates the greatest number of variants and proposes them manually as he calculates them. Eventually, by exclusion, one of them will be correct.

One of the applications that we expect to have from the development of *"Quantum Physics"* is precisely this: the generation of random sequences that allow you to generate passwords that guarantee the total inviolability of your data: at the same time we would also have discovered the algorithm behind the apparent randomness of the motion of the elements of the Universe and explain the relationship.

A universal law should be followed which brings together all the data already collected.

There would seem to be an explanation for everything, fluke shouldn't exist.

Albert Einstein, who defended this way of understanding the laws that govern the Cosmos, expressed his acceptance of a causal world by telling physicists who defended indeterminism that.

> *God is subtle but not malicious*

And he was certain that the Universe

> *hides its secrets not because it deceives us, but because it is essentially sublime*

For him there was certainly an intrinsic causality in the laws governing Nature, it is we who cannot see them.

In view of this, any phenomenon occurring in nature should not be due to the case. But then, is the probabilistic interpretation of *"Quantum Mechanics"* causal or random?

If the same scientists do not follow the rules that they themselves establish because they need the imagined result, rather than the one obtained by adapting the solutions from time to time according to necessity, then, from that moment on, is it true or is it not true that an untested rule becomes a carrier for a theoretical implant?

The Standard Model, perhaps, is based on this initial doubt and works until it produces the results I expect, if the results are not what I want the Model is no longer good for me and I start again looking for another hypothesis that makes it work better, But the means used to square the theory always remain unchanged and so we can't figure out if it's a definitive theory, or if it's valid just because we've made so many adjustments.

The problem is that we will never find any quantum physicist who can admit the use of this research method. Instead, a physicist, precisely for the definition of the term, should always seek a certain explanation of the phenomena and this should not be the result of the fluke.

And we should make sure of that.

Assuming this concept to be true, we would find it easier to admit that there are things we do not know how to explain, such as the initial singularity, the tunnel effect, the entanglement, the Huygens pendulums, etc., rather than ignore them or let the explanation itself become a non-truth that harms those who are unable to grasp the difference. Of so many events we know they happen, we know how they happen, but almost always, not why they happen.

An example?

The initial singularity from which, perhaps, everything originated: after the initial moment we built a *Celestial*

Mechanics, a *"Quantum Mechanics"* and continue to perfect the understanding of these ideas.

From the true beginning, let it be really clear once and for all, we, common men and scientists, know absolutely nothing: I have never heard a scientist of any discipline say what happened at that initial moment.

It is precisely because of this lack of knowledge that we cannot exclude anything.

However, many people deny a priori a creative act, and all this seems absurd to me.

Partial judgement should not enter the cultural baggage of a scientist.

Initially, all possibilities, equal to factors, should be equally likely to be considered.

This, in my opinion, does a disservice to science, because from that moment on it is no longer objective, but it becomes so if it suits me.

Or, and in this I open myself to other possibilities often overlooked, there are other theories that we have wrongly set aside.

We try to get out of the fog of our stereotypical knowledge: there are events that, when they occur, we do not understand and we set aside because they would undermine our beliefs.

The main question is what we have already asked ourselves about the role of the observer and which now comes back to the main question: does the experimenter, with his role as an observer, condition the final result?

Is the observer, with his personal interpretation, able to objectively do what he saw during the experiment?

Or did he just see what his limitations allowed him to see?

Or did the experiment adapt to the observer by showing him what he expected?

In *"Quantum Physics"*, there are no definite answers to these questions. It could certainly be said that these are theoretical investigations, but the double slit experiment will always be an example of a system that acquires the presence of an observer.

But if that were true, is it the observer who somehow determines the result, or would we have obtained the same result in any case?

But if the observer determines the result, does that mean that, in some way that we do not know, the system has been conscious of the observer and has produced a result that he expected to see?

If you want to take this thought to the extreme, you might think that things happen regardless of our will, or it is we who, in a way that we do not know, influence the course of events so that they happen in a certain way. Or do they just happen to imagine them without a real will to happen?

Is this not the explanation of what intuition can be?

Someone will have heard of *Synchronicity*, which is a synthesis of previous remarks.

Synchronicity, to simplify it is every time you say "Uhhh, what a coincidence..."

It happens every day: you arrive at a crossroads where no one passes by and the moment you arrive, just in that moment, there comes only one car that forces you to wait until it has passed and there are no more. What a coincidence.

Or you dream of a trip to the Bahamas and you start to find the Bahamas everywhere, on TV, in random articles, on TV shows. Uhhhh, what a coincidence.

Synchronicity, which means happening together in time, is all those events that seem to happen randomly but instead force you to think it's because of your destiny.

Synchronicity is a way of sensing the Universe as a whole, a single mechanism of which you are also a part, and which you believe that with your very existence, you provoke a series of events that may seem to be the result of coincidences.

Or Synchronicity could be a way of bending the results of an observation or event to the observer's expectations.

Reality would be built in the course of work.

It would be a shocking conclusion if it were possible to prove this hypothesis...

It would mean that *"Quantum Physics"*, like any other theory, became such because some of the scientists began to have the expectation that some events would happen that way.

This would explain why unusual discoveries, such as Planck's constant for example, still not explained mathematically, have solved many problems that cannot be solved otherwise.

Even the explanation of *entanglement*, still deeply unknown, would fall within those phenomena either wanted or completely divorced from our knowledge.

Maybe it's the answer to a desire to be able to make interstellar travel.

We have seen in the course of the events we have described how many discoveries have occurred by fluke and, we have called this kind of event with the name of Serendipity.

I am sure that many will prefer to think that they are all the fruit of imagination and that we have come to express philosophical concepts. Yet this is not the case.

The great Swiss psychoanalyst Carl Gustav Jung described an encounter with one of his patients in which he used a coincidence to improve his psychological condition.

The patient told him about a dream in which she received a jewel in the shape of a golden scarab.

At that moment Jung felt something banging against the window pane: unbelievable but true, it was a golden green bug resembling a beetle!

Seizing the coincidence, she opened the window and took the beetle: at that moment it was what he needed, but at the same time it was what she, the patient, needed.

And by handing her the insect he had established a connection between her dreams and objective reality.

This allowed Jung to resolve the lady's illness.

But do you think that the lucky one was Jung, or the lady who was able to heal her nightmares through that beetle?

This is the question to ask.

Jung, struck by the event, began to take an increasing interest in the topic, assigning three categories of coincidences similar to synchronicity:

- The first coincidence is that between a psychic content, such as a thought or a dream, a desire and an external event that occurs in the surrounding environment.
  This is a coincidence in time and space.
  For example, while you think about someone, you see them coming
- The second category is that between a psychic content and an event that occurs at a distance in space and, therefore, does not concern your life but is the result of your conscious or unconscious mind.
  This is a distant coincidence in time and space.
  The classic example is the premonitory dreams of catastrophic events.
  These cases are very frequent, just think of those who have had the premonition of an earthquake or some other tragic event.
- The third category is that between a psychic content and a coincidence that does not coincide temporally with a corresponding

external event but that will occur over time, perhaps after years.

For example, that any incurable disease is defeated, which for the moment is a dream, an unfulfilled hope, but which could occur over time.

In 1952 Carl Gustav Jung published a book called *Synchronicity as a Principle of Acausal Connections* and guess who gave him advice and contributed to the scientific part of the writing of the book? Wolfgang Pauli. Yes, the Nobel laureate physicist, one of the pillars of *"Quantum Physics"*.

All this will seem strange, but *"Quantum Physics"*, evidently, takes perception to the limit and pushes us to think that, in conclusion, the events that happen to us throughout our life are intensely connected with our unconscious.

The story of the friendship between Jung and Pauli is narrated in the book *"The equation of the soul"* by Arthur Miller: the physicist was one of the pillars of the newborn *"Quantum Physics"* and transformed the certainties of classical mechanics into clouds of probability.

But the scientist who daily confronted Bohr and the other great physics, at nightfall, turned into an unrepentant womanizer, often drunk.

Depressed and drunk, he often wandered in the red light districts of his city and mingled with prostitutes and drunkards.

Because of this double behavior he turned to Jung, who was particularly attracted to this type of pathology.

This meeting led to the development of a common language between Physics and Psychology, a kind of link between Matter and Spirit.

According to Miller, Jung contributed to Pauli's statements, especially to the formulation of the *Exclusion Principle*, through the schematization of his archetypal conception and his treatment of numbers and symbols according to logics partially elaborated by geometry and theology.

For Jung, coincidence is important for the sense it has in those who live the experience.

The connection between our inner world and reality is the result of a well-determined logic, which has its foundations in a sedimented patrimony of archetypes or models internalized over the past centuries and which peoples have in common; this is commonly called *the collective unconscious* to which we and our psyche are connected and from which we derive meanings.

When a significant coincidence occurs, a Synchronicity, in our daily experience, it is because in that moment we need to complete ourselves and in some way our psyche "summons" it.

Louis Pasteur, without knowing it, had sensed something like this and had expressed it in a thought:

Fluke favors only
the prepared mind

I also believe that the creation of coincidence favors the prepared mind.

About Pauli and the inexplicable relationships with his work in Theoretical Physics, anecdotes are reported on very strange phenomena, known as the "Pauli Effect", attributed to him: sometimes when he entered a laboratory the devices inexplicably suffered anomalies.

One day an expensive measuring instrument that was in the Physics laboratory of the University of Gottingen was inexplicably damaged and ceased to function.

The director of the Institute contacted Pauli in Zurich to warn him of the incident, jokingly citing that it was due to the Pauli effect.

Following subsequent checks it was discovered that just at the moment of the breakdown, Pauli, who was returning to Zurich from Copenhagen, was right at the Gottingen station waiting to take a connection.

George Gamow in the book *Thirty Years That Shook Physics*, argues that the stronger the effect, the more talented the physicist.

In 1950 Pauli was at the Princeton cyclotron which caught fire.

This led him to verify if the accident was not really due to the effect that had his name and, following these events, he published an article entitled "Background-Physics", "Physics of the Unconscious", in which he investigated if there were relationships between physics and deep psychology.

And here we are not talking about any quack, we are referring to a Nobel Prize, father of *"Quantum Physics"*.

In light of these events, many colleagues, objective and realistic physicists, distrusted Pauli, so much so that his friend and scientist Otto Stern always prevented him from entering his Institute.

In the end, many agreed that the Pauli Effect can be recognized as a "synchronic phenomenon."

If you tell these stories to a physicist in general but above all to a theorist, whom you may have met occasionally, be prepared to defend yourself, both verbally and physically: none of them will ever accept such a story as reliable. Also confirming on this occasion that,

I want to hear what I want to hear, not what you have to tell me.

Then it becomes legitimate to ask ourselves in which direction we are going. Also because we had already seen that another great of science, Bohm, raised similar doubts.

The young Bohm, while he was a researcher at the University of California at Berkeley and working on plasma[22], unexpectedly verified that the electrons, when they became part of the plasma, assumed an interconnected behavior, abandoning the typical behavior of single particles.

Later, years later, he will tell how the whole of the electrons conveyed to him a feeling of "alive".

He also pointed out difficulties in accepting Copenhagen's interpretation: the quantum world could not be just indecision and chance, and subatomic particles could not be revealed only in observation and measurement.

---

[22] Plasma: gas containing a high density of electrons and positive ions

These conclusions led him to feel something deeper than the random madness of nature in the subatomic world.

Fearing the criticism of quantum physicists, Bohm sent Einstein his conclusions and he welcomed them with interest. For six months they continued to exchange correspondence, agreeing that we would need a deeper knowledge of what was happening in the world of *quantum*.

He received the same welcome from Feynman, who met in Rio de Janeiro after being removed from Princeton because of his desire not to be a snitch.

Its two highly innovative scientific theories on the causal interpretation of *"Quantum Physics"* and the theory of the order involved and of indivisible totality will remain pillars of physics.

Bohm has bequeathed to us a vision of the Universe, of Science and of the *One and Total Thought*, so if we want to give humanity a deeper knowledge of nature, we should stop dividing mathematical problems and models into parts that we find more easily solvable; instead, we must begin to have a vision together.

The conclusion of David Bohm is that the Universe is "One", paradoxically in accordance with what has been said in ancient Eastern philosophies and beyond. Bohm's assertions are supplemented by a description of the phenomenon of *entanglement* that unites and connects all the particles of the Universe.

According to the scientist, in this universe of all one, completely interconnected, space and time no longer exist separately but unite together mind and matter in reality.

In conclusion, *Synchronicity* should be a system that the Universe uses to make us understand that we are part of a single interconnected reality, both physically and mentally, i.e. not directly interacting: this would explain the *tunnel effect* and the *entanglement* and, that every macroscopic or subatomic event always has a reason as long as you want to see it.

There are many other aspects that, because of the doubts of our minds, we cannot address without prejudice: they concern the relations between religion, science and free will.

Everyone can choose how and if to stand, but one thing we have to be sure of: Physics, in its main protagonists, even in the light of so many unexplainable evidence, should engage on a broad spectrum much more than it does at present, having the courage, where other assumptions are found, to reconvene and review laws and theories that seem to work.

Alternatively, there will no longer be a Physics made by researchers and genius, but only ever more sophisticated technology, based on old theories and, we will no longer have *annus mirabilis*.

"Quantum Physics" is Creationist

It is absolutely incredible how the evidence we have just talked about has been proposed even in historical times when, essentially, physics did not exist, when it was based only on the mental speculation that some enlightened could understand.

In 1700's George Berkeley, an Irish philosopher and theologian, appeared to have had some insights, in formidable advance, into what would happen in the 20th century, claiming that *Esse est percipi, the being is the being-perceived*, that is: the object consists only in its being perceived.

For him, the material substances that we see and that constitute our reality simply do not exist: a table, a chair, exist only because there is a mind and the world around us is the set of ideas that in order to be considered existing need a will inspired by a motion, which for many can be divine, that perceives them.

The world, in all its consistency of things and events, is intuitive only thanks to the ideas spread about the world by its Creator.

These ideas are humanly intellectualized, becoming perceptible.

Simply put, Berkeley says that things only exist in our minds because they're spirit, so things are just ideas or will; reality is the result of a series of sensory stimuli directly perfused by our Creator.

At the same time, in the 700's, German philosopher Immanuel Kant, by sinking his knowledge into Plato's "noumeno", claimed that what we know of reality is what appears, rather than his being.

Two centuries later, *"Quantum Mechanics"* seems to follow the same ideas: reality does not exist until we make an observation. If we want to push ourselves to the limits of this idea, it is something that is not yet built, based on something that is probably there and that we can only detect if it becomes the realization of the will of the minds of the scientists.

If *Copenhagen Interpretation* was true, so that the subatomic events measured by us exist only because the observer is present, it would confirm this much earlier philosophical thought.

In summary, probabilistic phenomena, precisely because of their intrinsic nature, can give even more experimental confirmation only when someone manifests to measure the experiment.

They are conclusions that will never be accepted by a scientific community that defends with drawn sword, paradoxically given the probabilistic nature of the whole construct, the ideas of orthodoxy of a thought that faithfully follows rules that unfortunately are based on approximations.

And it is ironic to note that many fans of this branch of physics are openly against creationism, but using principles that would immediately reveal to an objective mind the creationist origins of the proposer, paradoxically unaware of himself.

Now I wonder why a theory has gradually arrived that strongly recalls these philosophical suppositions: the answers can in my opinion be two.

The first: it is a coincidence and, therefore, *"Quantum Physics"* should have developed anyway as soon as a genius arrived who had intuited it.

The second: it is a chronological sequence that unconsciously led the philosophical thinking of Berkeley and Kant to influence the future founding fathers of *"Quantum Physics"*, more or less consciously.

Anyway, for everything we have read so far, I wonder how a quantum physicist, a cosmologist and a scientist in general, can ignore a possible creationist motion.

You may not believe out of conviction or stubbornness and, I can also accept this, but I just can't accept those who use science to deny the existence of a Creator.

Indeed, it is science that admits in all circumstances that it is unable to explain the zero moment, the beginning of the entire Universe, if we can imagine what Universe is.

At the very least, one should still accept the idea of a Higher Being.

Albert Einstein, although in a period of his life he professed atheism, always had a spirituality that accompanied him as background noise. Speaking of nature, he said that he felt

> *the sublime imprint and the admirable order that are revealed in nature and in the world of thought.*

And he continued to believe in a God, even if different from that of the Bible.

> *I believe in Spinoza's God who reveals himself in the orderly harmony of what exists, not in a God who concerns himself with fates and actions of human beings*

A true God, not what we selfishly want and desire.

Einstein's beliefs derive directly from the thought of Baruch Spinoza, a Dutch 1600s philosopher, according to which we are a universal whole that interacts with a continuous becoming.

The discovery of *entanglement*, again for what little we have understood, if it is true that it undermines the locality, one of the cornerstones of classical physics, it is also true that it would seem to confirm that there is a natural order in all things due to an entity that permeates all the elements of reality and arranges them with a mathematical order.

Thus Nature coincides with the essence of the Creator.

Quantum Consciousness

If *"Quantum Theory"* tries to explain why and how matter exists, it is also a good basis for understanding the brain and his consciousness.

Those who did it, proposed three types of approach: the first thinks that consciousness is an expression of the quantum processes of the brain; the second believes that through quantum processes it is possible to understand consciousness, without necessarily referring to brain activity; the third, considers that matter and consciousness are substrates of an underlying reality.

The study of these quantum aspects of consciousness has led specialists to believe that *complementarity*, *entanglement*, *dispersive states* and *non-Boolean logic* are of considerable importance in mental processes.

These researches carried out quietly by some neurologists pioneers in these new disciplines, at a first approach, seem to highlight that the brain activity referable to those mental processes can have an explanation directly from *"Quantum Physics"* logic, thus reconnecting to very previous philosophical and religious currents of thought much earlier than the start of 20th century studies. And, no matter how sceptical you may be, the relationship between the entire Universe, observation experiments and consciousness have produced a series of results, which to call them strange is an understatement.

Let us remember that Penrose, another Nobel laureate, had even imagined that our brain is endowed with something, a structure undetectable by our knowledge, capable of influencing quantum experiments.

If this were true, the mechanism of communication between neurons and the element of the experiment would have to be investigated?

Daily "Quantum Physics"

> *Science has no purpose, unlike engineering research. Our greatest progress is due to scientists who did not aim for utility but for fun, curiosity, the desire to understand.*
>
> Richard Feynman

"Quantum Physics" has made everyone's life much more comfortable.

Every day we have to deal with events related to it and we unconsciously use equipment that makes, more or less directly, use the principles that govern it.

Chad Orzel in his book *"Breakfast with Einstein"* describes many everyday objects that, from the moment we start the day, influence our life, which is governed by the manifestations of the strangest physical theory that has been developed.

For example:

The toaster: it is one of the most common objects that reveal its quantum properties. The internal resistance that toasts the bread, becoming incandescent until it changes color, emits sub-particles with a high energy state that modify the atoms of the bread surface.

Transistors: they are electron components that have become so common that they can be used in any electronic equipment: radio, TV, hairdryer, electric oven and thousands of other devices. The transistor is made up of layers of silicon and other elements which, when properly excited, transmit and modulate electrical signals. This excitement is a direct application of **"Quantum Physics"**.

Even all light sources, **neon, LED, incandescent bulbs**, etc., exploit quantum dynamics.

But surely the **Global Positioning System (GPS)** is the one that has brought the greatest benefits: today reaching and orienting ourselves in any place, even if unknown, from being a dream has become reality thanks to the help of *"Quantum*

Physics". Distance and time are calculated by connecting to satellites equipped with atomic clocks that are based only on *"Quantum Physics"*.

The **LASER**: the theory of laser operation is completely the result obtained from the studies on *"Quantum Physics"*. An excited electron jumps to a next energy level; eliminating the cause of the excitement, simply returning to its natural orbit and, by doing it, it releases the energy received in the form of light that we, with special devices, make it coherent, so to become a ray. The laser beam.

Possible Future Goals

> *Just because "Quantum Mechanics" is strange doesn't mean that all that is strange is "Quantum Mechanics".*
>
> *Victor J. Stenger*

Quantum Computers

Really there would be so many other examples to cite, but instead of looking at what has already been achieved, let us see what the benefits of *"Quantum Physics"* could be in the future.

Large companies such as Google, IBM and Microsoft, among the most well-known, but also private organizations and probably others not well known, are engaged in a competition to develop Quantum Computer, mainly for specialist applications.

This is a project whose realisation is very ambitious because of its intrinsic capacity of probabilistic calculations, intended to solve simultaneously multiple scenarios that would otherwise take years, if not centuries, to be solved with traditional instruments.

Despite the capabilities of the companies involved, and precisely because of their hardware architecture, they will not cancel the use of traditional computers, which will remain the simplest and cheapest solution to meet most of the needs of ordinary users.

The power of a quantum computer derives from its ability to work with quantum bits, or qubit.

Current computers use bits, a signal optical or electrical, which is 1 or 0.

This signal is the basis of any result a computer provides, be it a smartphone, a tablet, a laptop, any object has a computing ability.

So, photos, tweets, emails, songs and so on are just long series, organized according to well established codes, of these two values.

Quantum computers use qubits instead of bits, and they're subatomic particles like electrons or photons. We have seen that these particles have to respond to a much disciplined organization in order to exist and have many states instead of just two bits.

This multiplicity of capacities, called overlapping, of having multiple states and being able to do so simultaneously due to their probabilistic nature, increases their computing power more than exponentially.

The transition between the current computing power and the expected power will not be as from 1 to 100, but as from 1 to 1000000000, that is one billion times more powerful.

An astonishing change that will serve to make such progress that it is unimaginable at the moment.

Qubits have bizarre quantum properties, but some we can now understand, such as overlapping and entanglement.

Although we use these capabilities, this applies especially to *entanglement*, no one really knows how or why it works like this. It is precisely this unknown aspect that makes it so powerful, but it is at the same time its weakness, at least until we can understand why certain processes are taking place.

It is no coincidence that, if the power of calculation increases, there are more errors in quantum computers than in the computers we use every day, precisely because of their quantum nature that includes a phenomenon called *de-coherence*.

It is as if at some point a quantum computer decides to start telling lies. This happens because the qubits, interacting with the surrounding environment, decay until they disappear.

This is what we call *de-coherence*.

Basically, today a quantum computer is extremely delicate and subject to the whims of the environment in which it is immersed and this is making their progress problematic.

The race for *Quantum Supremacy* is however in full swing.

Quantum Supremacy will occur when a quantum computer unequivocally surpasses the computational capacity of a classical computer but, as there is no mathematical definition, it is difficult to determine when this will actually happen.

To date, Google has claimed this result but it has been contested by other competitors.

And let's not forget that we do not know what is happening in the laboratories of all those organizations that do not want publicity but are still working without communicating the results achieved.

Quantum Cryptography

There is another great milestone that now seems to have been achieved: talking and explaining quantum cryptography seems complex because it probably is.

It is one of the most recent results obtained thanks to *"Quantum Physics"* and further studies are proceeding rapidly so that we can achieve a protection system for all our data, as if they were in an inviolable safe.

Cryptography is a process that converts numeric or letter data in such a way that it can only be understood by someone who has the right "key" to read it.

Quantum cryptography, by extension, uses the principles and elements of *"Quantum Mechanics"* to encrypt data, transmit it, and then decrypt it downstream in an inviolable record.

Contrary to the descriptive simplicity, the realization of the process is quite complex precisely because of the mechanics that govern *"Quantum Physics"*.

There are many criticalities: the state of the particles is intrinsically uncertain since they can exist in several places at the same time, photons are not uniquely determinable, quantum properties, if measured, change; you can partially clone the properties of a particle but not the whole.

Hence, the great limitations for quantum cryptography.

Antimatter Space Propulsion

> *Behind me rang the Commander's voice: ignition engines and maximum power for warp speed.*
>
> *In absolute darkness the dim lights of the stars reappeared, but we did not recognize even one!!!*

This could be what's going to happen in the future in a man-made starship room.

Today, together with teleportation and time travel, this technology is purely theoretical, feasible on paper but without any practical idea of possible realization.

Science fiction draws it to its fullest and the stories and movies are realized so well, precisely because they are based on established physical theories, so much so that many people confuse a fantasy movie with a documentary.

For the time being, it must be clear, there is no such thing, but scientists are working on those foundations that could lead to further development in the future.

In particular, they are working on the theoretical development of an engine that uses the matter-antimatter reaction as a source of energy to use new engines that allow us to cross our Universe of belonging and explore it according to our needs.

With current knowledge, the laws of physics prevent us from thinking that there is anything that can surpass the speed of light, but with the development of an engine like this, we would get so close that we could achieve goals other than reaching a satellite or a planet close to us.

A matter-antimatter engine would bring the stars considerably closer outside our solar system and expand new extrasolar horizons.

How would an antimatter engine work?

Astronomers have found that, diffused in the Universe, there is something else that they think can be antimatter, in less quantity than matter. But there is and is the opposite of matter.

In the 1928's, Paul Dirac made corrections to the famous $E=mc^2$ considering the mass not only as positive but negative also. This allowed the existence of antiparticles that were then found experimentally. The antiparticle is exactly the same of matter but with inverted electrical charges.

In 20th century we found positrons that are electrons with negative charge, antiprotons, even their protons but with opposite charge and therefore negative charge; Using these two sub-elements at CERN, thanks to the high energy available, scientists built an anti-atom, in fact they built hydrogen antiatoms that remained alive for about forty nanoseconds.

The close contact between antimatter and matter causes the annihilation of the two particles with the production of pure radiation that escapes at the speed of light and produces other subatomic particles such as debris. Scientists speculate that this energy produced is the most powerful that can be generated.

Despite this knowledge, we have not yet built an engine based on this technology, because there is no antimatter available to us next in the cosmos. We are sure of this because we do not see any glow around us that would indicate the annihilation of particles.

Scientists have discovered a possible accumulation of antimatter in the center of the galaxy. If this is confirmed, we should look for a way to achieve it, but for now we are forced to produce it ourselves with the accelerators available.

NASA scientists believe they can build an engine based on this technology and have calculated that it would take just one millionth of a gram to produce enough energy to take a spaceship from Earth to Mars.

There will be no waste in the matter-antimatter propulsion, because all fuel, the annihilation of particles, is converted into energy. So the engines will have the chance to take us anywhere with very little fuel.

Amenities

> *Science tells us the how, but not the why. We know how it is made and what happened during the Big Bang, which was the beginning or a phase of the Universe, but we cannot know why there is matter, because from a soup of atoms and particles we arrived at the elements, at the stars and at living beings. For science these are facts, but there are no explanations.*
>
> *Margherita Hack*

Having Protons in the Head

To date it is not yet a way of saying, but it could also become one:

> *Leave me alone: today I have protons in my head ...*

In reality, as absurd as it may seem, it really happened.

The last place, certainly not recommended, in which to go and poke our heads is in a particles accelerator. Not even the need to dry your hair or a sudden itch is a good justification: it is, in fact, one of the worst ideas that can cross your mind and, probably, one of the last things you could do.

Yet on July 13, 1978, this is exactly what Anatoli Bugorski, a thirty six years old assistant at the Institute of High Energy Physics in Protvino, a city south of Moscow, did, casually, while working for his PhD at a proto-synchrotron known as U-70, at that time one of the largest accelerators in Russia and in the world.

During a normal working day, while checking for a trivial malfunction, a safety system failed to work and Bugorski, who had put himself on the path where the particles were conveyed, was hit by an accelerated proton beam with about 70 GeV.

The beam entered Bugorski's skull from the left back and exited his nose after passing through a part of his brain. In a few seconds 200 thousand rads were absorbed[23]: according to the

---

[23] the rad, replaced by gray in the International System, is the unit of

measurements common to all research methods, a dose greater than 600 rad is sufficient to kill a person

He was hospitalized and the effects of radiation were observed. According to many scientists, he was supposed to die within three weeks at the most.

Within twenty-four hours, Bugorski's face swelled up so much that it was unrecognizable, and skin and hair detached near the entry and exit points of the beam. The damaged skin, bones and brain tissue allowed the path of the beam to be traced. The scientist, however, luckily survived without cognitive damage, completed his research doctorate and continued to work as a researcher.

Strange consequences occurred in the next two years: the left side of his face gradually paralyzed while the nerves continued to destroy themselves. This event prevented the aging of the affected part of the face and today it still appears as it was at the time of the accident in 1978.

He has lost hearing from his left ear and has been the victim of occasional tiredness and convulsions: these are the only lasting effects.

It is still under control today, because it appears to be a more unique case that is rare to examine. Bugorski said that when he was crossed by the beam of protons, he saw a flash "brighter than a thousand suns", but without feeling pain.

For the amount of radiation absorbed, in any other situation Bugorski would have died.

---

measurement for the amount of energy absorbed and retained by the irradiated matter. (WikiPedia source)

However, no one until then, or later, has ever experienced such amounts of radiation in the form of a proton beam traveling close to the speed of light through the face.

This incident was discovered more than ten years later, due to the secrecy of the Soviet regime and its rules of confidentiality which prevented anyone who was present from speaking.

What conclusions can be drawn from this incredible event?

The radiation emitted by the particle beam should have killed Bugorski within a few days of the event due to the destruction of chemical bonds in the DNA of the affected cells, as happened in the victims of the Chernobyl disaster eight years later. But there was a substantial difference: in Bugorski's case, the radiation was concentrated only in the beam that had crossed his head and probably missed hitting any vital part of his brain. These coincidences, combined with the formidable ability of the brain to restore connections and functions when limited and limited damage occurs, have avoided an infamous outcome.

Some scientists have also suggested another hypothesis, referring to the concept of *Bragg peak*, which is the theory behind *Adronic Therapy*.

According to this hypothesis, the heavy charged particles, such as the protons in our case, but also the carbon ions, both adrons, from which the term adrotherapy, when they pass through a dense material, such as Bugorski's face, release energy, progressively increasing until they reach the peak at the end of the path, causing damage only shortly before the path of the beam comes to an end.

Presumably, according to this hypothesis, in the case of Bugorski the beam would have developed the peak only on the face without affecting the brain

Paradoxes

In science, paradoxes have often been used to explain some theories or to refute them.

But what do we mean by this concept and how can we use it?

A paradox allows, through a statement contrary to logic and common sense, to demonstrate a truth or a logic of the statement itself.

Paradoxes are often real thought experiments which, if solved, contribute significantly to affirming or denying theories.

Normally, because of their construction and their function, they are fun and become the subject of debate.

Paradoxes apply not only in physics but also in many other fields of application and always play the same role.

We have already seen and explained two of the most famous paradoxes, the Einstein-Podolsky-Rosen Paradox, or EPR Paradox, and Schrödinger's Cat Paradox, but there are many others:

> *The paradox of the two glasses of water is very simple: if we half-fill two glasses of water and place them in separate rooms with two observers, one will describe it as half empty and the other as half full. The paradox is that they are both lying but both are telling the truth.*

> *Another very famous paradox is that of the Grandfather and is based on the impossibility of traveling in time without changing events: suppose we have invented a time machine and travel until one of the grandparents is a young man who has yet to know his future grandmother. If, while walking in your car, you pass him and kill him, there is no chance that your father or mother exists because your grandmother will never know your grandfather. But if your father or mother does not exist, neither can you. But then what would the reality be?*

Another very famous paradox is that of the Twins and is a thought experiment that explains the effects of time dilation through the theory of relativity.

> *Two twins, named John and Joseph, turn twenty and decide to start working. John becomes an engineer and Joseph an astronaut. While John builds buildings, bridges and structures on Earth, Joseph is sent to the stars and travels the cosmos at close to the speed of light for about five years. When Joseph returns to Earth he is twenty-five years old.*
>
> *As soon as he returns, he goes to his brother John's house to greet him and finds a 95-year-old old man. But it's just John. What happened?*

The explanation is provided by the Theory of Relativity and the paradox allows us to understand how it works: the twins move differently in two different reference systems. Time flows for the two in a different way, faster for Joseph and in the usual way for John. This happens due to the *Lorentz Transformations* which are a standard part of the *Theory of Relativity*, so while for Joseph the journey lasted only 5 years, for his brother John who remained stationary on Earth, they corresponded to seventy-five years.

This paradox, known as the Gemini Paradox, is a scientific paradox, but also a logical one, and lends itself to a twofold interpretation.

The first, logical, makes me wonder how old is Joseph? Joseph lived for twenty-five years, but on Earth he is, like his brother John, ninety-five years old. So the right answer, albeit absurd, is that he is both twenty-five and ninety-five at the same time, it depends on how we measure age. There is no right or wrong age. Both are correct. The only one that could have something to say is his pension system which would not agree to assign him the pension at twenty-five.

The second interpretation is more physical and is the heart of relativity: for Joseph who travels at speeds similar to those of light, time has slowed down. From his point of view, the one who moves away is John and, therefore, considering things from his point of view, he should age up to ninety-five and John remain at twenty-five. Does relativity imply symmetry of points of view in these situations?

In reality, the explanation is inherent in the unfolding of the event: Joseph must accelerate to reach the speed of light while John is in a constant speed system, so the two points of view are not symmetrical since Joseph is the one who undergoes a significant acceleration and that's what the least time goes by.

Numerous very intriguing and often somewhat complex science fiction films in the development of the plot are based on this paradox: *The Bootstrap Paradox, the World War II Paradox, The Tolerance Paradox, the barber's paradox, etc.*

The paradoxes show that reality is not always able to provide a univocal answer, indeed it demonstrates that in many situations it is insufficient for its own evidence and the answer is uncertain even if the question and assertion are clear.

The use of paradoxes requires a good irony both in those who enunciate them and in those who listen to them: the ability to get involved and know how to resist the joke has been a characteristic of many great scientists.

The following paradox, which continues to raise questions and doubts in the best physicists of our time, I wanted to include last, because it deserves a more accurate discussion.

> *Fermi's paradox refers to a probabilistic calculation: since in the Universe and in our Galaxy there are billions of stars like the Sun, there will also be many Earth-like planets around some of those stars and, again for the calculation of probabilities, there should be many alien civilizations, many of which are as intelligent or more intelligent than us, so there should be many alien civilizations in the galaxy. So where are they, since we haven't met any yet?*

The raw statistics seem to support the paradox but, in my opinion, the chances of finding life that we can recognise are crumbling dramatically if we set some limits that seem obvious to me.

The various dating methods estimate that the Universe has an average age of about thirteen billion years and our Galaxy about ten billion years. We can deduce that there will be younger galaxies and older galaxies. Our solar system should be four, five billion years old and the first organism of which we have traced 3.5 billion years. The evolution of our species has taken this long time to produce technological intelligence. The combination of these data already makes it clear how difficult it is to achieve similar conditions simultaneously in the Universe. In addition, we must add other variables before we even consider the limits of our biology.

Another aspect to consider is the geological composition of the various celestial bodies: there will almost certainly be a diffusion of elementary organisms, which due to constraints imposed by the location are destined at best to remain such without technological development. Conversely, the biological limits of the same organisms are intended to be investigated in greater depth.

Let us begin with the intrinsic limits to our species: the duration of life and therefore of the civilization connected to it, and the ability to distinguish something different and any technology produced. The length of our lives prevents us from acknowledging that there are beings that do not correspond to our expectations: for example, if there were a sentient life form that is born and dies within the 1000 year period of our years, We won't recognize you and you wouldn't recognize us in the same relationship we have with plants or fruit flies.

We do not grant a sentient intellect to a plant, yet there is evidence of their ability to think from plants that we cannot understand.

Similarly, we do not believe that a fly can develop a technological world, especially because we have not studied the reason for it.

Other organisms, probably, do not evolve technologically because they do not have the time: I think of the octopus which, judging by their motor and intellectual abilities would be structurally capable of developing a civilization but do not progress because, simply, they live for too little time to be able to learn, assimilate and transfer knowledge.

So, we should find celestial bodies that host a set of civilizations that are compatible with our parameters, that essentially become one valid one: since the birth of the human species, Only in the last ninety years have we developed a technology capable of recognising other civilizations that have at least reached our technological level, which, for a purely temporal factor, I believe to be at the beginning and still at the beginning. Now, even if we overlap technologically, they should also coincide with the times of biological evolution. Analysis to be transferred to a universal scale of thirteen billion years in order to perfectly center 100 years of technology. I think it's easier to look for a needle in a haystack than a condition like that.

The alternative, at least for the level we have reached, which I believe is little more than nothing, is that there are technologically advanced civilizations that can interpret our stutter. Let me give you an example: we are sending probes into space to our neighbour and radio signals to deep space. If a civilization had evolved beyond that period of a hundred years, it could be at a stage where it is unable to recognise what we are sending, or it could be at a later stage. Perhaps they have developed the technology to transmit remotely using the properties of the entanglement and no longer use or have never used radio waves.

In view of these possibilities, the field of research is further reduced.

And we did not consider that we might not be able to recognize any intelligence because it was too different from us.

Fred Hoyle, the well-known English astronomer and physicist, in his "The Black Cloud", hypothesized an interstellar intelligence based on cloud-like agglomerations with completely alien intelligence, so much so that they are not mutually recognizable.

And if all the hypotheses on the table on the origin of the Universe were acceptable, how do we think we can not accept even the creationist hypotheses? They currently have the same level of probability as all the others. Or are we so stubbornly short-sighted that we don't want to accept such a possibility?

Let's not forget, I repeat for the umpteenth time, that no theory has so far explained the zero moment and no theory has excluded or confirmed the modalities. So why should we a priori deny a creationist motion?

Fred Hoyle himself put it that way:

> *Perhaps it is paradoxical. But isn't it even more paradoxical the idea that a lot of stuff, the entire Universe, was born in an instant, from nothing? (...) I find the idea of the creation of a hydrogen atom per year more acceptable than that of the birth of the Universe from one point.*

Anecdotes and Aphorisms

Among the most famous anecdotes and aphorisms, there are certainly those attributed to Einstein, which are also among the most iconic:

- "If you can't explain yourself, it just means you don't understand it yet."

- "The true distinguishing element of intelligence is not knowledge but imagination."

- "If it is proven that my theory of relativity is valid, Germany will say that I am German and France that I am a citizen of the world. If my theory is wrong, France will say that I am a German and Germany that I am a Jew."

- "When a man sits two hours in the company of a beautiful girl, it seems a minute has passed. But let him sit on a stove for a minute, and it will seem to him that two hours have passed; that is the relativity."

What I'm about to tell you is a joke about Einstein. Because of the multiplicity of calls generated on magazines or on the Internet, I am not sure that this actually happened, but it is so representative that it would seem a pity not to mention it:

"When Albert Einstein was still little known but his theories were now spreading, he was often invited to lecture around the United States.

However, his image was not yet in the public domain and few were able to recognize him. During one of those trips, his driver told him that he had attended so many of his lectures, that he now knew all his theories by heart and could easily speak in his place.

Einstein laughed at the thing that tipped him off for a fun find.

So, in a little-known place where they were headed, he decided to switch roles with his driver so that he, impersonating Einstein, would hold the conference in his place because he knew the contents of the report very well.

Upon arriving at the conference venue, Einstein let the driver go to the conference table and he sat in one of the last seats at the end of the great hall.

No one noticed the deception and the public believed they were facing an absolute genius.

The real Einstein watched the editing with tremendous fun, until someone in the audience asked a question to which the driver could not answer.

Not at all embarrassed and with great readiness of spirit he replied: "I wonder about you, this question is so obvious that even my driver would be able to answer... and then the scientist who responded as the driver surprised everyone."

Speaking of anecdotes or aphorisms in Physics certainly means speaking of Richard Feynman, a whimsical genius who loved life and who lived science very serenely, contending for the primacy of champion just to Einstein.

As already mentioned, in the evening it was easy to find him in Los Angeles bars sipping beer and playing his bongo, or like the time he encouraged Bohm in Rio de Janeiro, while he was spending a sabbatical year in that country that he had granted himself and that engaged in hunting for beautiful girls and beers.

Some of his expressions have become guides for entire classes of students:

- Always consider that your solution to a mathematical question has an understandable sense based on the question posed.

- Do everything possible to ensure that your communication is as unequivocal and frank as possible.

- Always be helpful and kind.

- Remember that life is an exciting adventure.

- Never lose your sense of humour.

- It's great if you can find something that you enjoy doing when you are young and that is big enough to keep your interest alive for the rest of your life. Because whatever it is, if you do it right (and so it will if you really like it), you will get paid to do what you would have liked to do anyway.

Curiosity

Among the most tantalizing curiosities is the one related to the number of electrons or atoms of which we are composed.

We start from a weight of a human body of 10 kg, so that we can easily multiply the result obtained by a quantity that relates it to our weight.

We must consider that our body has a varied composition of various elements, but with such percentages: oxygen 65%, carbon 18%, hydrogen (10%), nitrogen (3%), 4% other elements. By calculating the number of atoms per gram of each element we determine the number of electrons present for each element in 10 kg.

Approximately our 10kg body has almost 3.4×10^{27} electrons! Now remembering the weight of a single electron, the total weight of all electrons in the body, considered to be 10kg, is about 6 grams; a weight which, however small, is in proportion a value which, referring to a probabilistic quantity, shows a determined value. Mysteries of "Quantum Physics".

Electrons are only 0.03% of body weight, which is mainly determined by protons and neutrons which are respectively 1836 times and 1839 times heavier than electrons.

Other ...

> *Our species is allowed to*
> *perceive only photons that pulsate*
> *in a modest range of wavelengths:*
> *how to observe the world from a slit.*
>
> *Lidia Sella*

Chaos

In the theories concerning *"Quantum Physics"* there are elements that at the moment do not clearly fall within the hypotheses that are formulated, but that are to be considered because they have a decisive role in all the phenomena that occur in our physical Universe.

Chaos is one of these: a system is considered chaotic if the solutions of the equations that represent it are heavily influenced by the initial conditions. If these change even slightly, the solutions are completely different and follow a completely different path.

Weather forecasts are a chaotic system, indeed from these forecasts we have been able to really understand that small local changes cause climate changes capable of upsetting the conditions of the planet.

Edward Norton Lorenz (1938-2008) was able to verify this thanks to simple mathematical models he had devised and which were translated into programs that ran on computers. In 1960 he realized that, by re-entering numeric strings with different decimal approximations, he obtained completely different projections with results that completely changed the final scenario. And this happened every time he even intervened, alone, with small variations.

Lorenz had verified that there are systems that have unpredictable modes of behavior in which a single variable, even if slightly modified, heavily influences the final history of the system. Chaotic systems are considered the fields of economics, aerodynamics, population biology, thermodynamics, chemistry and medicine, as well as particle physics and cosmology.

This is confirmation that the Universe, from *"Quantum Physics"* to macroscopic systems, has unforeseeable consequences due to both the numerous variables and the even minimal changes it has undergone.

Another factor that should always be taken into consideration is *entropy*.

Entropy

In the second half of the twentieth century, thermodynamics started talking about gradients, that is, the progressive differences of quantities such as temperature or pressure.

Gradients are the real engine of the essence of life: in fact nature tends to reduce gradients, but fortunately it is nature itself, stimulated by evolution, which opposes this trend. Energy tends to dissipate due to the second law of thermodynamics: this phenomenon is Entropy.

Entropy measures the degree of disorder of a system: the lower it is, the more orderly the system tends to be and the energy is maximum; the higher the entropy, the more disordered the system is and the energy is minimal.

The temperature called absolute zero, 0 degrees Kelvin, corresponds to -273.15 degrees Celsius. At that temperature, the molecules don't move. Everything is orderly, but there is no more life. That's why nature is opposed to reaching that temperature.

For living beings and for nature, thermodynamic equilibrium is equivalent to death.

The complexity of life is the tendency to reduce gradients and oppose the tendency of nature.

This is a way of understanding evolution: the understanding of gradients increases with the awareness of our ability as a single individual to react to environmental stimuli; improving intelligence understood as knowledge of the system to which one is part, pushes towards a tendency to optimize our abilities.

When social improvements and intelligence become grouped, different stimuli come into play linked to the synergies between the individuals in the community. In this way a group intelligence develops that no longer have the brakes and the quality of the intelligence of the individual and activates even self-destructive processes that can lead to the exploitation of resources until they are exhausted.

In this we see another way of interpreting "Quantum Physics": the consciousness of an imperfect system, which however makes possible a group evolutionary trend towards a possible alternative solution, which is not said to be the best choice.

The error is used to correct the error, in a completely unconscious way.

Together with the two previous factors, it is as if there were many other possibilities that are scarcely considered but which play determining roles. Impossible to list them all, because it is also difficult to identify them.

Many Little Truths

Even Einstein was forced to artifice to make ends meet some of his hypotheses.

So to propose the model of the static Universe it was enough for him to modify only the equations on which general relativity is based, adding concepts that included a "cosmological constant".

For the *Theory of General Relativity*, space-time and gravity are the same physical concept: space-time curvature is the description of gravity.

It was not Lawrence, with the accelerator technique, who started the hunt for elementary particles using non-radioactive sources, but it is true that he invented the best technique to do so. Before him John D. Cockcroft (1897-1967) and Ernst TS Walton (1903-1995), in 1932 had caused the mutation of lithium atoms into two particles, obtaining the necessary energy from a potential difference. And something similar had also happened previously using the generators developed by Robert J. van de Graaff (1901-1967).

All particles have the corresponding antiparticle and when they meet they annihilate each other generating energy.

Hadrons are divided into two categories: baryons, protons, neutrons, hyperons and mesons, particles whose mass have values between those of an electron and a proton.

Gell-Mann (1995, 198) to whom we owe the discovery and naming of quarks in 1963, called them so based on a sound more like "cuorc". Later reading James Joyce's "The Vigil for

Finnegan" he found the word quark. This word is the sound of a seagull's chirp. Hence the use of the term for discovery.

The Transforming Neutrinos

In 2015 the Nobel Prize in Physics was awarded for a strange discovery. By now we have understood that in the subatomic world the term "strange" is almost a custom. Neutrinos were found to have the ability to change as they travel through space. Referring to the characteristic of "flavour" it is as if it changes depending on when and where it is tasted. This is a funny example that perfectly describes the strangeness of neutrinos.

The problem is that for this to happen the neutrino would have to have a non-zero mass, but this would go against the Standard Model of Particle Physics.

It has long been thought that neutrinos have no mass, like photons, in fact in the second half of the twentieth century it was discovered that they do.

To date this question poses many perplexities to scientists and one of the most fascinating researches is precisely that of understanding their strange abilities.

Other small truths, as I liked to define them, are two phenomena that still cannot be explained:

Hong-Ou-Mandel Effect

This phenomenon is little known but similar to entanglement, to tunnel effect and to particle interferometry.

It is a quantum optics phenomenon, which Chung Ki Hong, Zhe Yu Ou and Leonard Mandel discovered in 1987, which occurs when two particles of light strike a surface. Individually, the two photons would go half the time in one direction and the other half in the other. But if the two photons leave together, zzzzaaaaaccccc …… here is the magic!

Photons always travel together in one of two possible directions: it never happens that they travel in two different directions: it is as if there was only one driver at the wheel who makes decisions for everyone.

The other phenomenon, equally unknown to most, also remains without a rational explanation.

Better Than a Pinball Machine

In everyday reality if we have to go from one place to another place we know that, if there are two paths, we can only follow one.

In quantum reality it doesn't work like this: if we take a photon, when at a crossroads it has to choose whether to go right or left, it doesn't think twice and at the same time chooses the two paths. But the photon always remains one. This was seen in a recent experiment carried out in 2018.

The photon, instead of choosing, decides not to choose and goes through both paths at the same time. And it is impossible to establish even in retrospect if and which path he followed. We only know that it is at the end of the path.

This too is a phenomenon whose causes are still unknown.

Breaking news

Odderon e Muon g-2

The results of the experiments carried out over the years have, however, provided results that are well suited to the hypotheses formulated and, this confirms their authoritativeness, although there remain formidable doubts about the overall system.

The latest news coming from CERN and FermiLab confirm the same trend: new discoveries but also formidable doubts.

On March 16, 2021, just in these days as I am writing, a semi-particle theorized in the 1970s, the Odderon, seems to have been confirmed.

Timothy Raben, a particle theorist at Kansas University explained the significance of this event with an example:

> *It is as if the protons were two large car transporters that, carrying cars, collided. Cars fell apart on the road, falling out of car transporters, and the pieces got confused, making it impossible to determine which car they belonged to...*

My four year old grandson does the same thing with his cars: he collides them, retrieves the pieces and proudly shows them. As already reported in another chapter, nothing has changed with respect to the methods of the atomists who intended to divide matter as far as possible with the tools at their disposal: scissors or blades against accelerators.

Without going into excessive technicalities to explain the dynamics of the experiment, it can be said that the detector of these particles is so sensitive and sophisticated, that we cannot guarantee the true success of the experience, since what it has detected is so out of the ordinary observations which we do not know if it is a trace of something else or really the observation that was expected. In fact, the official statement speaks of "possible presence".

On April 8, 2021, another probable discovery was made: physicists from Chicago's Fermilab announced an anomaly in the behavior of muons, in an experiment called Muon g-2.

The meaning of this event is not the verification of something that we expect to find, but the attempt to measure a deviation of a value from what we know by which we could understand what we do not yet know.

I prefer that the conclusions on these two experiments are drawn directly from the readers, who by now will have knowledge of the system.

In both cases, in my opinion, it is clear that we are dealing with scientific sensationalism, probably attributable to the desire to draw attention or to journalistic exaggerations by rethinking the results obtained. Thus the non-discovery at the expense of certainty is more evident in both cases.

Thus, a bad method that has become customary is perpetuated. Science, precisely because of the difficulties in explaining the new possible goals, does not need a media megaphone. I find the behavior of scientists who studied, experimented and published the results only in journals that validated the results before publishing them much more elegant...

The habit of turning the lack of denial or certainty into an enlightened theory, which the profane embrace willing to crusades convinced of their goodness, is now invaded.

In this way, only science that becomes a victim of itself pays: imagine a scientist who, convinced of the truthfulness of the non-verifiable results of certain experiments, develops, pending final verification, a theory from which he gets other answers that have little to do with the original questions of the experiment. From that moment on, a long research phase begins and it checks out virtually nothing.

In my opinion, an alarm has been sounded.

I will never finish repeating that *"Quantum Physics"* seems to have become a scientific discipline for a few elected people, and that, having made great progress in the first part of the last century, it has been scraping the bottom of the barrel for decades. I repeat, even the most recent acclaimed discoveries such as Higgs Boson and Gravity Waves are discoveries that come from the past rather than the future.

Physicists are at the pinnacle of intellectual knowledge and contempt and, for this reason alone, they have to realize that, furious at waiting for the future, that has already become the past.

Reading news such as these of the last two discoveries, I have the same doubts of Sabine Hossenfelder, who works at the Institute of Advanced Studies in Frankfurt and, I wonder if they are really a symptom of scientific progress and, above all, whether they are functional to technological development for an evolving civilization.

Perhaps we should start studying observers; obviously done in a scientific way.

The scientific laboratories have evolved from small cellars, huts or, in any case, from limited areas, where with few means available, an experimental nature was sought as far as possible perceptible until it reached gigantic clusters of increasingly sophisticated and usually exclusive researchers and machines in their circle.

They, I hope unintentionally, form interconnected clusters on specificities and form genuine neural networks where each ganglion is the research group that decides what to share.

By examining, observing and evaluating behavioural systems through specially designed algorithms, it would be possible to verify with predictive methods individual and group behaviours aimed at research.

The observers themselves would become part of the experiment and I think that the behaviour in their closed system would be influenced by the external observers who are examining them...

Perhaps scientists have become wizards, and they know that a trick has three parts:

- *what people see;*
- *what he remembers;*
- *and what he will tell others he has seen, and I hope that one day we will not realize that even for science,*

as Schiller said

> *"Even the gods fight in vain against stupid people"*.

In any case, I trust in "Stein's law":

> *if something cannot go on forever, sooner or later it will stop.*

If there were still doubts, or someone thought my observations too vehement, I would suggest reading Karl Popper's *"Quantum Theory and the Schism in Physics"*, which explicitly mentions the irremediable rift that divides physicists between "orthodox" and "rebels" ".

A way of saying that there is no one side that is more right than the other, as it seems obvious to the critics of the system that someone is wrong. And for Popper it would be necessary for someone to collaborate to understand what, where and when the mistake could have been made.

Modello Standard

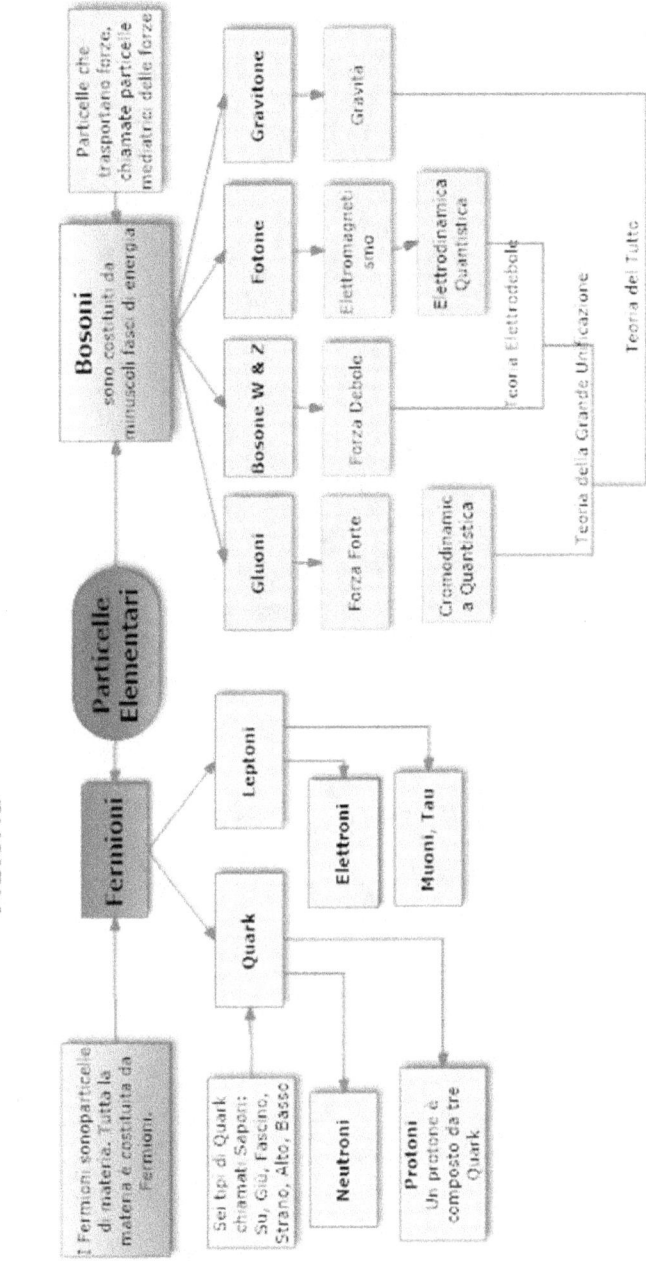

Appendix

History of the Electron

Corpuscle	Wave
1833 M. Faraday hypothesizes in electrolysis the existence of a fundamental atom of electricity. The term atom, in this context refers to the old meaning of the term, that is, the smallest part of matter without any other reference to its elementary components.	
1869 W. Hittorf applies an electric discharge in gases and discovers cathode rays.	
1874 G. J. Stoney estimates the value of the elementary charge.	
1881 H. Helmholtz re-proposes Faraday's hypothesis on the electricity atom.	
1891 G. J. Stoney calls the atom of electricity "electron"	

1897 J. J. Thomson associates cathode rays with negatively charged particles by calculating their mass / charge ratio, and discovers that they are a thousand times lighter than the hydrogen atom.	
1899 J. J. Thomson finds that the electron has a negative charge.	
1913 R. A. Millikan perfects the measurement of the electron charge	
	1923 L. de Broglie hypothesizes that the wave-particle dualism proposed by Einstein for the photon also holds for particles of non-zero mass.
	1928 G. P. Thomson, son of J. J. Thomson, confirms de Broglie's theory.
	1957 J. Faget and C. Fert carry out electron interference both with

	Fresnel's biprism and repeating Young's two-slit experience

1976 P. G. Merli, G. F. Missiroli and G. Pozzi directly observe the statistical process of the formation of interference fringes thanks to an electron microscope.

1989 Prof. A. Tonomura at the helm of a group of Japanese Hitachi researchers replicates the experiment of P. G. Merli, G. F. Missiroli and G. Pozzi using a more refined instrumentation that takes advantage of the 13 years of technological progress since 1976.

Main Particles Timeline

Discovery year	Particle	Discoverer
1885	He discovers that the atoms emit positive radiation which Rutherford will call the Proton	Eugen Goldstein
1897	Electron	Thomson
1911	Nucleus of the Atom	Rutherford
1920	Proton	Rutherford
1920	Neutron	Chadwick
1922	Photon	Arthur Compton
1932	Positron antielectron The first	Carl D. Anderson from Paul Dirac hypothesis during

	antiparticle discovery	1927 and Ettore Majorana in 1928
1937	Muone or mu lepton	Seth Neddermeyer, Carl D. Anderson, J.C. Street, and E.C. Stevenson
1947	Pion	C. F. Powell's from Hideki Yukawa hypothesis in 1935
1947	Kaone	G.D. Rochester e C.C. Butler
1955	Antiproton	Owen Chamberlain, Emilio Segrè, Clyde Wiegand e Thomas Ypsilantis
1956	Neutrino	Frederick Reines e Clyde Cowan from Wolfgang Pauli studies in 1931
1979	Gluone	observed

		indirectly in the events at DESY
1983	Bosons W e Z	Carlo Rubbia, Simon van der su studi di Sheldon Glashow, Abdus Salam e Steven Weinberg
1995	Quark top	Fermilab
2012	Higgs Bosone	CERN

These are not all the particles discovered but only the main ones are reported to give an idea of the evolution of "Quantum Physics" and the Standard Model.

Bibliography

Author	Title
Aspect, A., J. Dalibard e G. Roger	*Test sperimentale delle disuguaglianze di Bell utilizzando analizzatori variabili nel tempo*
Atti Royal Society	*Presenza di singolarità in cosmologia.*
Bell JS	*Sul paradosso di Einstein-Podolsky-Rosen*
Clauser, JF, MA Horne, A. Shimony e RA Holt	*Esperimento proposto per testare le teorie delle variabili nascoste locali*
Crease Robert P.	*Il prisma e il pendolo: i dieci più bei esperimenti scientifici*
Einstein Albert, Podolski Boris e Rosen Nathan	*La descrizione quantica della realtà può essere considerata completa?*
Georgi, H. e SL Glashow.	*Unità di tutte le forze delle particelle elementari.*
Gilder Louisa	*The Age of Entanglement*
Greene Brian	*L'Universo elegante*
Greene Brian	*La trama del cosmo*

Hawking Stephen	*Presenza di singolarità in universi aperti*
Hoyle Fred	*Un nuovo modello per l'Universo in espansione*
Kuhn Thomas	*The Structure of Scientific Revolutions*
Landau, L.	*Sulla teoria delle stelle*
Linde, AD	*Un nuovo scenario dell'Universo inflazionistico*
Miller Arthur	*L'equazione dell'anima*
Miller Arthur	*In Equilibrio perfetto*
Penrose Roger	*La strada che porta alla realtà*
Robinson Andrew	*Thomas Young: The Man Who Knew Everything*
Robinson Andrew	*L'ultimo uomo che sapeva tutto*
Schweber, S.	*Una prospettiva storica sull'ascesa del modello standard*
Toraldo di Francia G.	*L'indagine del mondo fisico*
Venema Liesbeth	*Light, enchanted*
Weinberg Steven	*I primi tre minuti: una visione moderna dell'origine dell'Universo*

Weisstein Eric W.	*Il mondo della scienza di Eric Weisstein*
Witten, E.	*Dinamica della teoria delle stringhe in varie dimensioni*

Summary

Presentation 5

My Grandma 11

We Were Only Men 15

The Universe..21

Theory Satisfaction....................................25

Atoms..29

First Atom Definition 31

Infinite Idea's, Origin and End of All........37

Atomic Theory Is a Good Idea 41

1800 First Atomic Theory of Matter 43

1880 Atomic Structure Time 45

The Big Race 57

1900 The Universal Exposition in Paris 59

1904 The Plum Pudding Atom Model 65

"Quantum Physics" 69

   Planck: Quantum Physic's Dad 71

   Scientific World Reaction's 79

   Investigations on the Nature of Light 83

   Young's experiment: the most beautiful 86

Einstein Grandmother 89

   Introduction to Relativity 91

   Special relativity 97

   General relativity 99

   Photoelectric effect 107

The Atomic Models 109

The Handover 111

1911 Rutherford Father of Nuclear Physics 115

1913 Bohr: Einstein's Opponent and Friend 127

Bohr Heirs 131

1919 The Proton 133

Particles Exist … 137

1923: The Compton Effect 139

1924: Louis de Broglie and the Duality Wave .. 141

1925 Pauli Exclusion Principle 145

1925 Heisenberg and the Matrices 149

1926 The Compton Photons 153

… or They Probably Exist? 155

1925 Thomas Kuhn and The Events onwards ... 157

1926 Schrödinger and Born 161

1927 Heisenberg and his Principle 168

1928 Dirac 170

October 29, 1927: Fifth Solvay Conference and Copenhagen Interpretation 175

Einstein Bohr Duel 180

Wave Function Collapse 186

"Quantum Physics" Comes of Age 193

1929 First Particle Accelerator 195

Roaring Twenties 196

Physics Goes to War 201

Heisenberg Cold Case 203

The Neutron 213

1930 Gödel 214

1934 Panisperna Street Boys 221

1935 Schrödinger's Paradox 231

   The New Proposal 233

   The Observer in Quantum Reality 235

   1935 Schrödinger's Cat 240

We can never be innocent again 245

Moe Berg: the Probable Heisenberg Killer 247

August 6, 1945: Hiroshima 254

Atomic Bomb Explosion: the day after 256

"Quantum Physics" From 1950 onwards 259

The World of Physics after the War 261

EPR Paradox, Bohm and the Hidden Variables 262

Bohm's Hidden Variables 268

Alain Aspect Hot Shot 274

1982 Alain Aspect 275

1985 Giancarlo Ghirardi, Alberto Rimini and Tullio Weber 277

The Particle Accelerators 283

What They Are and How They Work 285

CERN e LHC 289

The Last 50 Years 297

The Standard Model 299

What is missing? 304

Higgs Boson's Story 306

Quarks 308

String Theory 309

Big Bang and Particle Physics 313

2021 Has Physics Lost the Right Road? 323

Madness or a New Path: Double Slit Experiment
.... 325

The Quantum World 329

The Crossroads 332

Are There Other Ways? 335

A Possible Solution 337

The risks 339

The Education System 342

    No, It Is Not True 347

    A New Scientific Policy 348

    At the End...349

... Nothing Has Changed: We Are Only Men! 353

Side Effects of "Quantum Theory"... 359

    Case, Coincidence, Synchronicity, Serendipity.361

    "Quantum Physics" is Creationist 377

    Quantum Consciousness....................................381

Daily "Quantum Physics".......................................383

Possible Future Goals 387

    Quantum Computers..389

    Quantum Cryptography 392

    Antimatter Space Propulsion 393

Amenities 397

    Having Protons in the Head 399

    Paradoxes 403

    Anecdotes and Aphorisms 411

    Curiosity 415

Other … 417

    Chaos 419

    Entropy 421

    Many Little Truths 423

    The Transforming Neutrinos 425

    Hong-Ou-Mandel Effect 426

    Better Than a Pinball Machine 427

Breaking news 429

Odderon e Muon g-2..431

Appendix 439

History of the Electron 441

Main Particles Timeline.....................................445

www.ingramcontent.com/pod-product-compliance
Lightning Source LLC
Chambersburg PA
CBHW071443220526
45472CB00003B/645